CLINICAL GENETICS IN BRITAIN: ORIGINS AND DEVELOPMENT

The transcript of a Witness Seminar held by the Wellcome Trust Centre for the History of Medicine at UCL, London, on 23 September 2008

Edited by P S Harper, L A Reynolds and E M Tansey

Volume 39   2010

First published by the Wellcome Trust Centre
for the History of Medicine at UCL, 2010

The Wellcome Trust Centre for the History of Medicine
at UCL is funded by the Wellcome Trust, which is
a registered charity, no. 210183.

ISBN 978 085484 127 1

All volumes are freely available online following the links to Publications/Wellcome Witnesses at
www.ucl.ac.uk/histmed

CONTENTS

ILLUSTRATIONS AND CREDITS

ABBREVIATIONS

ACC	Association of Clinical Cytologists
AFP	alpha-fetoprotein
AGNC	Association of Genetic Nurses and Counsellors
BSHG	British Society for Human Genetics
CEGEN	Confidential Enquiry into Genetic disorders by non-geneticists
CF	cystic fibrosis
CGS	Clinical Genetics Society
CMO	Chief Medical Officer
CPK	creatine phosphokinase
DoH	Department of Health
DHSS	Department of Health and Social Security (1968–88)
DMD	Duchenne muscular dystrophy
GCRB	Genetic Counsellor Registration Board
GIG	Genetics Interest Group
GNSWA	Genetic Nurses and Social Workers Association
GOS	Great Ormond Street Hospital for Children NHS Trust
GPSI	general practitioner with special interest
GTAC	Gene Therapy Advisory Committee
hCG	human chorionic gonadotrophin
HFEA	Human Fertilization and Embryology Authority
HLA	Human Leukocyte Antigen
HPC	Health Professions Council
JCHMT	Joint Committee for Higher Medical Training
MRC	Medical Research Council

NHS	National Health Service
NIHR	National Institute for Health Research, Leeds
PGGS	Primary Care Genetics Society
PKU	phenylketonuria
RCGP	Royal College of General Practitioners
RCP	Royal College of Physicians of London
SMA	spinal muscular atrophy
SMD	special medical development
UCH	University College Hospital, London
UCL	University College London
WHO	World Health Organization
WNB	Welsh Nursing Board

WITNESS SEMINARS:
MEETINGS AND PUBLICATIONS [1]

In 1990 the Wellcome Trust created a History of Twentieth Century Medicine Group, associated with the Academic Unit of the Wellcome Institute for the History of Medicine, to bring together clinicians, scientists, historians and others interested in contemporary medical history. Among a number of other initiatives the format of Witness Seminars, used by the Institute of Contemporary British History to address issues of recent political history, was adopted, to promote interaction between these different groups, to emphasize the potential benefits of working jointly, and to encourage the creation and deposit of archival sources for present and future use. In June 1999 the Governors of the Wellcome Trust decided that it would be appropriate for the Academic Unit to enjoy a more formal academic affiliation and turned the Unit into the Wellcome Trust Centre for the History of Medicine at UCL from 1 October 2000. The Wellcome Trust continues to fund the Witness Seminar programme via its support for the Centre.

The Witness Seminar is a particularly specialized form of oral history, where several people associated with a particular set of circumstances or events are invited to come together to discuss, debate, and agree or disagree about their memories. To date, the History of Twentieth Century Medicine Group has held more than 50 such meetings, most of which have been published, as listed on pages xiii–xvii.

Subjects are usually proposed by, or through, members of the Programme Committee of the Group, which includes professional historians of medicine, practising scientists and clinicians, and once an appropriate topic has been agreed, suitable participants are identified and invited. This inevitably leads to further contacts, and more suggestions of people to invite. As the organization of the meeting progresses, a flexible outline plan for the meeting is devised, usually with assistance from the meeting's chairman, and some participants are invited to 'set the ball rolling' on particular themes, by speaking for a short period to initiate and stimulate further discussion.

Each meeting is fully recorded, the tapes are transcribed and the unedited transcript is immediately sent to every participant. Each is asked to check his or her own contributions and to provide brief biographical details. The editors

---

[1] The following text also appears in the 'Introduction' to recent volumes of *Wellcome Witnesses to Twentieth Century Medicine* published by the Wellcome Trust and the Wellcome Trust Centre for the History of Medicine at UCL.

turn the transcript into readable text, and participants' minor corrections and comments are incorporated into that text, while biographical and bibliographical details are added as footnotes, as are more substantial comments and additional material provided by participants. The final scripts are then sent to every contributor, accompanied by forms assigning copyright to the Wellcome Trust. Copies of all additional correspondence received during the editorial process are deposited with the records of each meeting in archives and manuscripts, Wellcome Library, London.

As with all our meetings, we hope that even if the precise details of some of the technical sections are not clear to the non-specialist, the sense and significance of the events will be understandable. Our aim is for the volumes that emerge from these meetings to inform those with a general interest in the history of modern medicine and medical science; to provide historians with new insights, fresh material for study, and further themes for research; and to emphasize to the participants that events of the recent past, of their own working lives, are of proper and necessary concern to historians.

ACKNOWLEDGEMENTS

'Clinical Genetics in Britain: Origins and development' was suggested as a suitable topic for a Witness Seminar by Professor Peter Harper, who assisted us in planning the meeting. We are very grateful to him for that input, to Professor Martin Bobrow for his excellent chairing of the occasion and to Professor Sir John Bell for writing the Introduction to the published proceedings. We thank Professors Angus Clarke, Peter Harper, Ursula Mittwoch, Bernadette Modell, Heather Skirton, Sir David Weatherall and Dr Caroline Berry for their help with the Glossary and Professors Joy Delhanty, Dian Donnai, Alan Emery, Peter Harper, Shirley Hodgson, Marcus Pembrey, Sue Povey, Dr Nick Dennis and Ms Dallas Swallow for help with photographs and graphs. For permission to reproduce Figures 6, 8 and 12 included here, we thank medical illustration, Institute of Child Health and the paediatric research unit, Guy's Hospital, London.

As with all our meetings, we depend a great deal on staff at the Wellcome Trust to ensure their smooth running: especially the Audiovisual Department, Catering, Reception, Security and Wellcome Images; Mr Akio Morishima, who has supervised the design and production of this volume; our indexer, Ms Liza Furnival; and our readers, Ms Fiona Plowman, Mrs Sarah Beanland and Mr Simon Reynolds; and Ms Stefania Crowther for editorial and marketing assistance. Mrs Debra Gee is our transcriber, and Mrs Wendy Kutner and Ms Stefania Crowther assisted us in running this meeting. Finally we thank the Wellcome Trust for supporting this programme.

Tilli Tansey

Lois Reynolds

Wellcome Trust Centre for the History of Medicine at UCL

VOLUMES IN THIS SERIES*

1. **Technology transfer in Britain: The case of monoclonal antibodies**
   Self and non-self: A history of autoimmunity
   Endogenous opiates
   The Committee on Safety of Drugs (1997)
   ISBN 1 86983 579 4

2. **Making the human body transparent: The impact of NMR and MRI**
   Research in general practice
   Drugs in psychiatric practice
   The MRC Common Cold Unit (1998)
   ISBN 1 86983 539 5

3. **Early heart transplant surgery in the UK (1999)***
   ISBN 1 84129 007 6

4. **Haemophilia: Recent history of clinical management (1999)**
   ISBN 1 84129 008 4

5. **Looking at the unborn: Historical aspects of**
   obstetric ultrasound (2000)
   ISBN 1 84129 011 4

6. **Post penicillin antibiotics: From acceptance to resistance? (2000)**
   ISBN 1 84129 012 2

7. **Clinical research in Britain, 1950–1980 (2000)***
   ISBN 1 84129 016 5

8. **Intestinal absorption (2000)***
   ISBN 1 84129 017 3

9. **Neonatal intensive care (2001)**
   ISBN 0 85484 076 1

*Volumes in print and freely available from Dr Carole Reeves, see page xvi.

22. **The Rhesus factor and disease prevention (2004)**
ISBN 978 0 85484 099 1

23. **The recent history of platelets in thrombosis and other disorders (2005)**
ISBN 978 0 85484 103 5

24. **Short-course chemotherapy for tuberculosis (2005)**
ISBN 978 0 85484 104 2

25. **Prenatal corticosteroids for reducing morbidity and mortality after preterm birth (2005)**
ISBN 978 0 85484 102 8

26. **Public health in the 1980s and 1990s: Decline and rise? (2006)**
ISBN 978 0 85484 106 6

27. **Cholesterol, atherosclerosis and coronary disease in the UK, 1950–2000 (2006)**
ISBN 978 0 85484 107 3

28. **Development of physics applied to medicine in the UK, 1945–1990 (2006)**
ISBN 978 0 85484 108 0

29. **Early development of total hip replacement (2007)**
ISBN 978 0 85484 111 0

30. **The discovery, use and impact of platinum salts as chemotherapy agents for cancer (2007)**
ISBN 978 0 85484 112 7

31. **Medical ethics education in Britain, 1963–1993 (2007)**
ISBN 978 0 85484 113 4

32. **Superbugs and superdrugs: A history of MRSA (2008)**
ISBN 978 0 85484 114 1

33. **Clinical pharmacology in the UK, *c.* 1950–2000: Influences and institutions (2008)**
    ISBN 978 0 85484 117 2

34. **Clinical pharmacology in the UK, *c.* 1950–2000: Industry and regulation (2008)**
    ISBN 978 0 85484 118 9

35. **The resurgence of breastfeeding, 1975–2000 (2009)**
    ISBN 978 0 85484 119 6

36. **The development of sports medicine in twentieth-century Britain (2009)**
    ISBN 978 0 85484 121 9

37. **History of dialysis, *c.* 1950–1980 (2009)**
    ISBN 978 0 85484 122 6

38. **History of cervical cancer and the role of the human papillomavirus, 1960–2000 (2009)**
    ISBN 978 0 85484 123 3

39. **Clinical genetics in Britain: Origins and development (2010)**
    ISBN 978 0 85484 127 1 (this volume)

40. **The medicalization of cannabis (2010)**
    ISBN 978 0 85484 129 5 (in press)

All volumes are freely available online at www.ucl.ac.uk/histmed following the links to Publications/Wellcome Witnesses.

*Volumes freely available, while stocks last, from Dr Carole Reeves at: c.reeves@ucl.ac.uk

Hard copies of volumes 21–40 can be ordered from www.amazon.co.uk; www.amazon.com; and all good booksellers for £6/$10 each plus postage, using the ISBN.

UNPUBLISHED WITNESS SEMINARS

1994   **The early history of renal transplantation**

1994   **Pneumoconiosis of coal workers**
(partially published in volume 13, *Population-based research in south Wales*)

1995   **Oral contraceptives**

2003   **Beyond the asylum: Anti-psychiatry and care in the community**

2003   **Thrombolysis**
(partially published in volume 27, *Cholesterol, atherosclerosis and coronary disease in the UK, 1950–2000*)

2007   **DNA fingerprinting**

The transcripts and records of all Witness Seminars are held in archives and manuscripts, Wellcome Library, London, at GC/253.

OTHER PUBLICATIONS

Technology transfer in Britain: The case of monoclonal antibodies
Tansey E M, Catterall P P. (1993) *Contemporary Record* **9**: 409–44.

Monoclonal antibodies: A witness seminar on contemporary medical history
Tansey E M, Catterall P P. (1994) *Medical History* **38**: 322–7.

Chronic pulmonary disease in South Wales coalmines: An eye-witness account of the MRC surveys (1937–42)
D'Arcy Hart P, edited and annotated by E M Tansey. (1998)
Social History of Medicine **11**: 459–68.

Ashes to Ashes – The history of smoking and health
Lock S P, Reynolds L A, Tansey E M. (eds) (1998) Amsterdam: Rodopi BV,
228pp. ISBN 90420 0396 0 (Hfl 125) (hardback). Reprinted 2003.

Witnessing medical history. An interview with Dr Rosemary Biggs
Professor Christine Lee and Dr Charles Rizza (interviewers). (1998)
Haemophilia **4**: 769–77.

Witnessing the Witnesses: Pitfalls and potentials of the Witness Seminar in twentieth century medicine
Tansey E M, in Doel R, Søderqvist T. (eds) (2006) *Writing Recent Science: The historiography of contemporary science, technology and medicine.* London: Routledge: 260–78.

The Witness Seminar technique in modern medical history
Tansey E M, in Cook H J, Bhattacharya S, Hardy A. (eds) (2008) *History of the Social Determinants of Health: Global Histories, Contemporary Debates.* London: Orient Longman: 279–95.

Today's medicine, tomorrow's medical history
Tansey E M, in Natvig J B, Swärd E T, Hem E. (eds) (2009) *Historier om helse (Histories about Health,* in Norwegian). Oslo: Journal of the Norwegian Medical Association : 166–73.

INTRODUCTION

As specialties go, clinical genetics has had a remarkably short history. Over the last 30–40 years, it has grown from an isolated activity to a fully integrated clinical activity bridging clinical and laboratory functions. The complexity of issues it has been asked to manage in a clinical setting has also expanded dramatically in this short time frame, while the clinical load of patients who benefit from increasing scientific insights into human genetics has grown. Because of its short history and its impact on all aspects of medicine, it is a perfect subject for a Wellcome Witness Seminar. Many of the original 'witnesses' to the birth of this specialty are still alive and provide fascinating insights into how a very few individuals shaped this important clinical service.

Since its early gestation in a few centres, notably the Galton Laboratory, the Oxford MRC population genetics unit and the clinical labs at Great Ormond Street, Manchester, Guy's and Newcastle, the specialty has fed off exciting advances in human genetics and applied these advances to improve human health. The early years were dominated by personalities such as Lionel Penrose, Alan Stevenson, John Fraser Roberts, Cedric Carter, Cyril Clarke, Paul Polani and others. This was an era when formal training was not established and the precise relationships between genetics and other disciplines were not defined. It was a time when the first laboratory function, cytogenetics, created the distinctive capabilities of the field that ultimately allowed the specialty to develop a unique position among the medical specialties. Clinical geneticists struggled with the longstanding associations with eugenics and were challenged by the limited opportunities to apply their growing science to patients, which eventually evolved to become a unique specialty, distinct but associated with established service functions in subjects such as paediatrics, neurology and general medicine.

The specialty confronted a range of challenges – recognition by the Royal College of Physicians and the Department of Health, the creation of an inclusive society to support its growth in the form of the British Society for Human Genetics, reconciliation between the laboratory and clinical functions of the service, and the need to evolve other paramedical capabilities such as genetic counselling. All these presented obstacles that were ultimately overcome as the isolated groups applying genetics to medicine came together as a professional body. Although small, the specialty has continued to have influence that far exceeds its size. This Witness Seminar clearly describes the formative years that created this entity, influenced throughout by the great international figures of clinical genetics such as Victor McCusick.

This story is crucial to a modern understanding of the challenges and opportunities confronting the specialty. The underpinning science has become one of the most important aspects of medical science and, through it, genetics has increasingly come of age, slowly but steadily infiltrating all specialties. The view of Cyril Clarke was particularly prescient, as it was he who encouraged the growth of genetics in general medicine and it is in this broader domain that clinical genetics is likely to eventually to end up. Genetics increasingly anchors our understanding of disease, and genetic diagnostics in the form of arrays and sequencing has now all but subsumed the laboratory functions that were developed 30 years ago. It seems unlikely that clinical genetics as it was originally conceived will be able to preserve its role in medicine. Molecular pathology is likely to assimilate the laboratory functions which will apply equally to haematology, paediatrics, neurology, cardiology and oncology. Pharmacogenetics is already becoming an important tool for stratifying medicine. The clinical domain of clinical genetics will not disappear but the role of genetics in medicine will mean that both laboratory and clinical functions will be mainstreamed in the future.

The crucial roles played by the actors in this seminar in introducing genetics to the clinic must be remembered as they pioneered many of the practical and ethical issues that are likely to be revisited repeatedly as genetics finds its place in a wider arena.

Sir John Bell
University of Oxford

CLINICAL GENETICS IN BRITAIN:
ORIGINS AND DEVELOPMENT

The transcript of a Witness Seminar held by the Wellcome Trust Centre
for the History of Medicine at UCL, London, on 23 September 2008

Edited by P S Harper, L A Reynolds and E M Tansey

CLINICAL GENETICS IN BRITAIN: ORIGINS AND DEVELOPMENT

Participants

Ms Chris Barnes
Dr Caroline Berry
Professor Martin Bobrow (chair)
Professor John Burn
Dr Ian Lister Cheese
Professor Angus Clarke
Dr Clare Davison
Professor Joy Delhanty
Dr Nick Dennis
Professor Dian Donnai
Professor Alan Emery
Professor George Fraser
Mrs Margaret Fraser Roberts
Professor Peter Harper

Dr Hilary Harris
Professor Rodney Harris
Professor Shirley Hodgson
Dr Alan Johnston
Mrs Ann Kershaw
Mrs Lauren Kerzin-Storrar
Professor Michael Laurence
Professor Ursula Mittwoch
Professor Michael Modell
Professor Marcus Pembrey
Professor Sue Povey
Professor Heather Skirton
Professor Tilli Tansey
Professor Sir David Weatherall

Among those attending the meeting: Sir Christopher Booth, Dr Michael Buttolph, Professor Bernadette Modell, Professor Oliver Penrose, Dr Maria Jesús Santesmases

Apologies include: Professor Michael Baraitser, Dr Cyril Chapman, Professor Michael Connor, Professor Peter Farndon, Professor Malcolm Ferguson-Smith, Dr Christine Garrett, Mrs Ann Hunt, Mr Alastair Kent, Dr Mary Lucas, Professor Norman Nevin, Professor Derek Roberts, Dr Joan Slack, Dr Mary Vowles, Dr Elspeth Williamson, Professor Ian Young

Professor Tilli Tansey: Good afternoon everyone and thank you very much for coming to this Witness Seminar on the history of clinical genetics in Britain. My name is Tilli Tansey and I'm the convenor of the History of Twentieth Century Medicine Group, which was established by the Wellcome Trust in 1990 to bring together scientists, clinicians, historians and others who are interested in recent medical history – conventionally post-Second World War medical history – to find out what happened, when, why, who the drivers were and who the dissidents in certain fields were, and to record these reminiscences and to provide records to go into archives for the use of present and future historians.

This meeting was suggested by Peter Harper, professor of medical genetics at Cardiff University, who has, in addition to being a geneticist, also transformed himself into an historian, and he has just produced a book of nearly 580 pages, called *A Short History of Medical Genetics*.[1] Of course, it's also very important to identify a chairman and we're very grateful that Martin Bobrow, emeritus professor of medical genetics at the University of Cambridge and former governor of the Wellcome Trust, has agreed to undertake this onerous task and to keep you all in order during this afternoon. So, without further ado, I'll hand over to Martin.

Professor Martin Bobrow: So, now you all know why I am here: it's to keep you in order. There certainly isn't any other obvious reason why I should get picked out for this signal honour. However, as someone who doesn't do history much, it is interesting to think about how clinical genetics over the time that I've known it – which I think is only a short time, because I've only just begun my career – and it has gone from next to nothing to a relatively mainstream bit of medicine, and is possibly on its way to becoming next to nothing again. I think it's interesting to consider the origins of the subject and the way that it has developed as a separate specialty.[2] And if it becomes next to nothing, it won't be because genetics becomes unimportant, it will be because it gets absorbed so much into the rest of medicine that it doesn't have as strong a claim to a separate existence. That is one of the issues and tensions that has been present in clinical genetics since its inception: most of medicine is divided into territorial specialties – liver people have the liver and knee people have the knee – but genetics doesn't quite have a territory; it stamps on everyone else's territory. It's not unique in that, but it is quite unusual and it leads to

---

[1] Harper (2008).

[2] For example, see Christie and Tansey (eds) (2006) on the development of the field of medical physics.

certain tensions in structuring. Its scientific roots are very complex, I think. It's easy to say: 'It's genetics', but genetics is a separate body of knowledge, which differs from most of the laboratory techniques, that are themselves diverse and complex, that underpin genetic practice. It has strong roots in epidemiology and statistics, in counselling, whatever you wish to define that as, but certainly in talking as a means of getting through problems. It has always had a very strong basis in population science, both in terms of thinking of populations, as opposed to patients away from populations, and in terms of looking for prevention and screening as opposed to just thinking of treatment, because this is a non-prescribing specialty, to note another unusual feature.

If one's interested in curiosities in medical practice, it is quite an interesting topic to play around with. Within the UK, most of the people who made the early running are sitting in this room now. There are a few people that I wish were with us today who aren't, some because they couldn't make it, and some because they aren't going to make any more meetings at all. But an awful lot of us, you, are sitting here. My job is just to keep you in order as we drift through a very loosely structured agenda (Table 1).

| **Origins and early development: the scientific roots of clinical genetics** |
| Lionel Penrose and the Galton Laboratory |
| Chromosomes and the need for clinical genetics |
| Alan Stevenson and the Oxford MRC unit |
| **The first medical geneticists** |
| John Fraser Roberts, Cedric Carter and the Institute of Child Health, London |
| Paul Polani and clinical genetics at Guy's Hospital, London |
| Cyril Clarke and the Liverpool Institute |
| Wider development of the field in Britain |
| **Bodies involved in developing the field** |
| Royal College of Physicians' Clinical Genetics Committee and the special advisory committee |
| Clinical Genetics Society and the British Society for Human Genetics |
| Department of Health |
| **Other aspects** |
| Dysmorphology and clinical genetics |
| Genetic counselling; the development of genetic nurses and genetic counsellors |
| The role of lay societies in the growth of genetic services |
| The ethical and social dimension in clinical genetics |

Table 1: Outline programme for 'Clinical Genetics in Britain: Origins and development' Witness Seminar

The way I am going to play it, and this is not deeply rehearsed and is subject to change, is that as we go down this loose agenda, I've scribbled a name and I will look at someone to start the topic rolling, after which anyone who has anything to say may chip in. There are quite a few of us; we are covering roughly 40 years of work and we've got about three and a half hours. So a bit of mental arithmetic will tell you that half-hour speeches are not what we need this afternoon. Brevity and a few facts, and things that are interesting and illuminating, which is exactly what you are all so good at. So, lastly, it is a real pleasure to see so many people who have been part of my working life, some of whom I haven't seen for years, and I'm looking forward to it.

Professor Peter Harper: Over the past few years I started to think a lot about how clinical genetics began and then developed in the UK. But there are a good few points that I'm still very uncertain about and I think this Witness Seminar gives a good chance to discuss, and hopefully clarify, these; even though, as Martin said, some of the people who might have contributed most are no longer living. Of course, there has been a previous Witness Seminar, seven years ago, on genetic testing, but this was focused mainly on laboratory aspects.[3] So this is the first opportunity in a group setting for looking at the equally important field of clinical genetics. You'll have seen that the first topic on today's programme is the scientific origin of clinical genetics, but I think we also need to think about the origins from different fields of medicine: paediatrics, adult general medicine and ophthalmology, as examples.

When did clinical genetics begin? I think most of us would probably put the date around the late 1950s when human genetics as a science was already reasonably well developed and human chromosome studies had provided medical genetics with a practical, laboratory basis, as well as a research basis. But, before the Second World War there were already people like Julia Bell (1879–1979) in the UK, and others like Madge Macklin (1893–1962) in the US and Canada, whose research was based on the combined clinical and genetic analysis of families.[4] I think that these and others like them should be regarded as forerunners of clinical geneticists, even though they did not actually practise

---

[3] The Witness Seminar was held in 2001 and published as Christie and Tansey (eds) (2003).

[4] See, for example, Bell and Haldane (1937); Macklin (1933). For an evaluation of Julia Bell's work, see Harper (2005); Bundey (1996); for details of human genetics in the US, see Reed (2005) and on medical genetics, see Leeming (2010) .

it as a medical specialty.[5] One such person who bridged these two periods was John Fraser Roberts and we're very fortunate to have his widow, Margaret, here to tell us something about him and his work today.

There's one uncomfortable question that we need to ask and that is whether clinical genetics developed to any significant extent from the earlier field of eugenics. I've always felt this not to be the case, at least in this country, but some people consider the two to be directly linked, so I think it's essential to look at how modern clinical genetics has developed what, in my view, is a very strong ethical dimension.

Once clinical genetics had become identifiable as a definite field of medicine, what were the factors that helped it to spread across the UK and become a flourishing specialty in its own right? Did the Department of Health and Social Security (DHSS) actively encourage it? Or was it the work and influence of just a few pioneers such as, for instance, Cedric Carter. After all, most of this development happened in the 1970s, when the economy wasn't a whole lot better than it is today. And, how did clinical genetics become so well established in the Royal College of Physicians, with its own trainees and specialty committee?[6] Then also how did it establish its own society, the Clinical Genetics Society (CGS)?[7] There are people here today who I think should throw some light on this.

The final aspect that I hope we'll be able to cover, although not nearly to the extent that it deserves, is the diversification of people working in clinical genetics, especially non-medical staff such as genetic nurses and genetic counsellors, and

---

[5] Professor George Fraser wrote: 'One of these forerunners was Lionel Penrose, from 1931 when he initiated the Colchester survey (Penrose (1938)) onwards for more than four decades until his death in 1972. He was the first clinical geneticist and, at the Galton Laboratory, he passed on his knowledge and wisdom.' Note on draft transcript, 28 February 2009; 2 March 2010.

[6] The Royal College of Physicians established the Clinical Genetics Committee in 1984 as a result of discussions in 1983 with Dr D A Pyke, the registrar of the college. Members were invited by the college, with Rodney Harris as the first chairman and Alan Johnston as honorary secretary. Among their reports are: Royal College of Physicians, Clinical Genetics Committee (1991a); Harper *et al.* (1996); Davies *et al.* (1998). See Appendix 1, page 84.

[7] The Clinical Genetics Society was established in 1970 to bring together doctors and other professionals who were involved in the care of individuals and families with genetic disorders. It is one of the founding constituent groups in 1996 of the British Society for Human Genetics, which provides a forum for all professionals involved in genetics as a clinical service and for research. See www.clingensoc.org and www.bshg.org.uk/society/about_us.htm (visited 7 January 2010).

the increasingly important role of lay societies in genetic developments. I think these are areas that would need a separate Witness Seminar to do them justice, but I'm glad we have some representatives here who can provide this kind of perspective. There is a project in progress trying to preserve and catalogue in detail the records of people in medical and clinical genetics, and Tim Powell, from the National Cataloguing Unit for the Archives of Contemporary Scientists in Bath, is coordinating this.[8] He's here at this meeting, so if any of you can think of key records that are buried somewhere that might be saveable, this is a good chance.

I think I've said enough to get this seminar started, but I'd just like to end with a couple of definitions that might help us to avoid confusion: by 'human genetics' I mean the scientific study of heredity in humans, both normal and as in genetic diseases; whereas by 'medical genetics' I mean the study of genetic diseases in their own right, including applications in medical practice; and I've taken 'clinical genetics' to mean that part of medical genetics involving the direct clinical study of, and communication with, patients and families themselves. These three areas overlap extensively, but the third one hasn't seen much attention in historical terms so far, so I hope that we can help to remedy this today.

Bobrow: The way Peter Harper has structured this – very helpfully, I think – is that you have three bullet points on your bit of paper (Table 1) on origins and early development, the first of which is Lionel Penrose and the Galton Laboratory at UCL.[9] We will go through those bullets one at a time just as discussion points with statements from anyone around who has anything to contribute on that topic. Before we launch into that, is there anything that Peter has said that people want to correct immediately, or violently disagree with? No, there's nobody twitching to that extent? Let's move on to Lionel Penrose (Figure 1) and the Galton Lab. I was hoping that Sue Povey would start.

Professor Sue Povey: I feel very embarrassed to be representing Penrose and the Galton Lab, because I only started work at the Galton Lab in 1970, though I did

---

[8] For further information on the National Cataloguing Unit for the Archives of Contemporary Scientists, see www.bath.ac.uk/ncuacs/home.htm (visited 17 February 2010). The unit was closed at the end of December 2009 as part of university economies and will be reconstituted as part of the Science Museum in 2010, but in the interim the Human Genetics Archiving Project is being continued at Cardiff University, funded by the Wellcome Trust.

[9] For further details about the work of the Galton Laboratory, see www.galtoninstitute.org.uk/Newsletters/GINL9112/Galton_Laboratory_Today.htm (visited 19 February 2010).

Figure 1: Professor Lionel Penrose, c. 1960.

visit it as a student and Penrose was there. And so, of course, being in the Galton Lab, one had a great awareness of Penrose's contribution to the understanding of tuberous sclerosis (epiloia) in 1935 and phenylketonuria (PKU) just after the war.[10] My personal experience of him was that he had great concern for the patients. As far as I understand it, he had very *ad hoc* consultations with patients. I don't think there was a standard genetic clinic at the time at UCL, but there are people here who were there at the time who could correct me on this. Penrose did once take me to see some Down syndrome twins who were supposed to be monozygotic, and one had Down's and one did not. What I remember most clearly is that he was so distressed by the terrible state of the high-rise building and the mother who couldn't get out. He was much more concerned with the welfare of the family than the science, which seemed to pale into insignificance compared with how this poor family were managing.

Regarding the naming of the Galton Laboratory, it is my understanding that he did that by changing his notepaper when he became Galton professor, after the war. In 1955 he tried to change the name of the department from the department

---

[10] See Gunther and Penrose (1935); Penrose (1946); see also Harris (1974); Christie and Tansey (eds) (2003): 35, 40, 41.

Figure 2: Dr Mary Lucas explains a poster to the University of London's
Chancellor, Princess Anne, October 1981.

of eugenics and the *Annals of Eugenics* to the *Annals of Human Genetics*, but UCL moved rather slowly, and it took about two years to do this, during which time he started using the name Galton Laboratory on his notepaper.[11] I may be corrected on that, but that is something that Bette Robson, Galton professor of eugenics at UCL (1978–93), said at the Penrose symposium I held for the centenary of his birth.[12]

My own memories of early clinical genetics, which you may well say was not early, are from 1967 when Dr Mary Lucas was appointed as a clinical geneticist at the Galton Laboratory (Figure 2),[13] and then the beginning of the MRC

[11]   The Francis Galton Laboratory of National Eugenics became the Galton Laboratory of UCL's department of human genetics and biometry in 1963 and part of the department of biology in 1996. The *Annals of Eugenics* was renamed the *Annals of Human Genetics* in September 1954 (volume 19 part 1). See Barnett (2004).

[12]   See Anon (1998). A detailed account of the Penrose symposium at UCL held on 12–13 March 1998 (Povey and Press (1998)), including a summary of the papers presented, is available on the department of genetics, environment and evolution website at www.ucl.ac.uk/gee/about/penrose_symposium. A list of sources of biographical information on Lionel Penrose provided by Professor George Fraser will be deposited, along with the other records of this meeting, in archives and manuscripts, Wellcome Library, London, at GC/253. For further details of the Galton Laboratory, see note 9. See also Laxova (1998).

[13]   See, for example, Lucas *et al.* (1972).

unit, when Gerald Corney (Figure 5) ran a clinic at which the notable things, I think, were that he accepted patients who referred themselves and always wrote to them afterwards, using great detail about what had happened and what the conversation had been. He did a great deal of background work on the diagnosis before ever agreeing to see the patients. I thought that I would mention Gerald, because people, perhaps, forget that particular clinic.[14]

Professor Ursula Mittwoch: I don't know the definition of a genetics clinic, but I do remember that every so often on a Saturday morning Lionel Penrose would see a child with Down syndrome and his or her parent, usually the mother, and he would advise them. He took palm prints of the child and the parents and I was called in to make blood films to study the polymorphonuclear leucocytes.[15]

Professor Joy Delhanty: I came to work at the Galton Laboratory to do my PhD in 1959. It was in 1959 when the first case of a chromosomal anomaly had been described and that was in a case of mongolism (Down syndrome).[16] My supervisor in genetics, John Maynard Smith, said he wondered whether Lionel Penrose would be interested, in view of that finding, in taking on a PhD student. That's how I came to work at the Galton Lab, and I went for a week to study with David Harnden at Harwell (MRC radiobiology unit, Atomic Energy Research Establishment) to learn the essence of tissue culture and how to grow skin biopsies.[17] I remember the excitement over the next couple of years, because Lionel was so interested in the families that he'd accumulated, particularly those where there were siblings with Down syndrome. In one of those, the early ones, we were able to demonstrate that there was an inheritance of a Robertsonian translocation, which was responsible for the recurrent Down syndrome. This was the first demonstration of that (Figure 3a and b).[18]

---

[14] See, for example, Crosse and Corney (1961). For biographical note, see page 117.

[15] Professor Ursula Mittwoch wrote: 'The leucocyte investigation was based on the finding by Turpin and Bernyer (1947) that the polymorphonuclear lobe count in patients with Down syndrome was lower than in unaffected individuals. The problem was that there were no controls for the patients visiting Professor Penrose. Dr Valerie Cowie kindly solved this problem by introducing me to the Fountain Hospital, Tooting, London, where the physician superintendent and staff were extremely helpful in selecting Down syndrome patients and appropriate controls, and facilitating the technical procedures. The results of the investigations on lobe counts and the incidence of drumsticks gave rise to several publications between 1955 and 1958. See for example Mittwoch (1958a and b).' E-mail to Professor Tilli Tansey, 29 September 2008.

[16] Lejeune *et al.* (1959); Jacobs *et al.* (1959); Harnden *et al.* (1960).

[17] See, for example, Harnden (1960, 1979, 1996).

[18] Penrose and Delhanty (1961); see also Penrose *et al.* (1960).

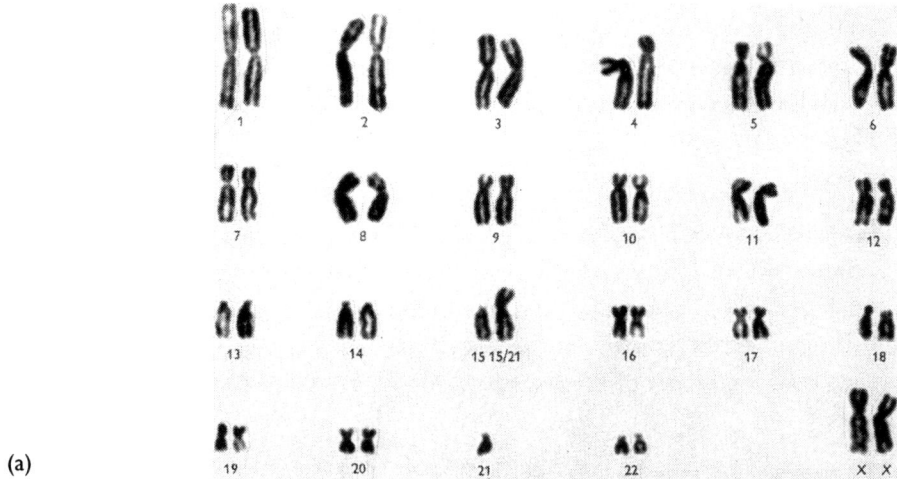

(a)

Figure 3(a): The karyotype of a phenotypically normal woman (no. 110 in 2nd row in b) who gave birth to four children (3rd row in b), two of whom had Langdon Down syndrome anomaly (nos. 160 and 183 in 3b), all of whom exhibit a fusion between chromosomes 15 and 21.

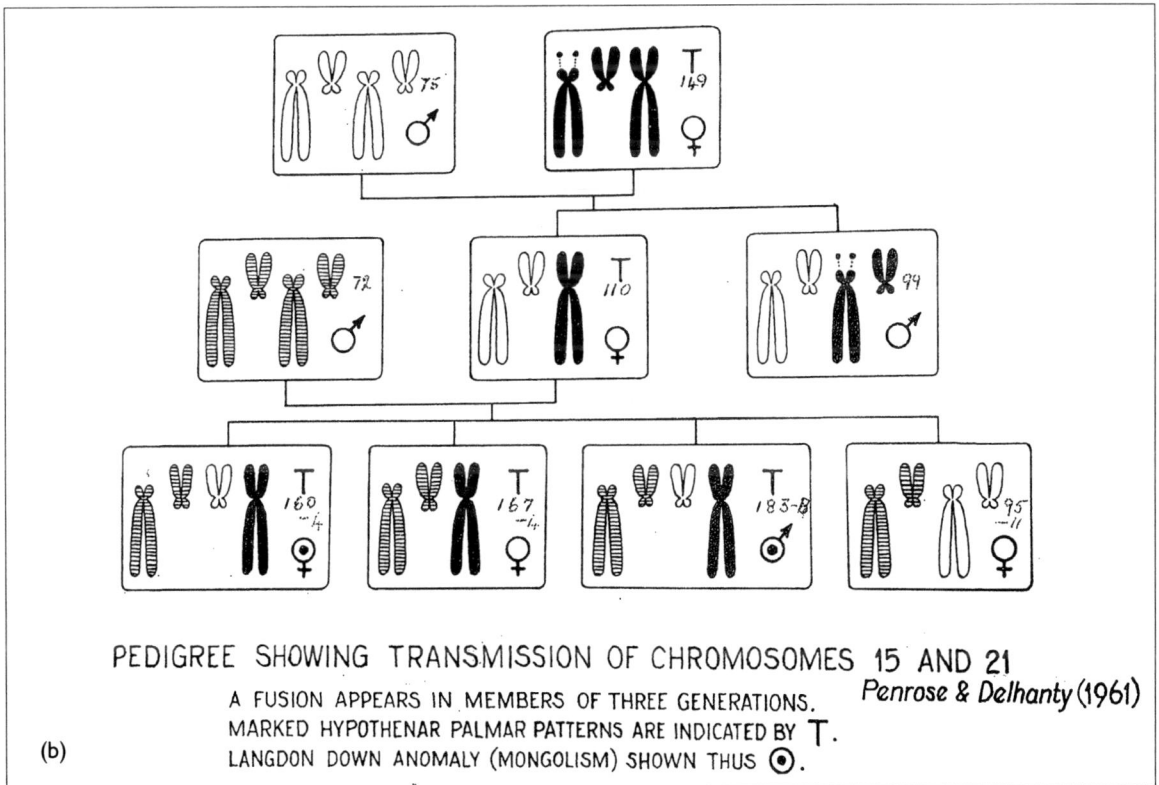

PEDIGREE SHOWING TRANSMISSION OF CHROMOSOMES 15 AND 21
A FUSION APPEARS IN MEMBERS OF THREE GENERATIONS. *Penrose & Delhanty (1961)*
MARKED HYPOTHENAR PALMAR PATTERNS ARE INDICATED BY T.
LANGDON DOWN ANOMALY (MONGOLISM) SHOWN THUS ⊙.

(b)

(b): Family pedigree, showing three generations with inherited translocation (marked with T). Penrose and Delhanty (1961).

He was so interested he couldn't keep away from the lab and the microscope. He was always coming and saying, 'What have you found today?' And sometimes it was to do with Down's and sometimes it was looking at miscarriages, and we did find the first two examples of triploidy in humans at that time. So it was a very exciting time.

Professor George Fraser: I should like to acknowledge my personal debt to Lionel Penrose, whom I first met 52 years ago and who was my teacher, my mentor and my friend. He was a pioneer in clinical genetics and medical genetics as well as in human genetics; in the years since our first meeting, I have not met anyone working in any of these overlapping fields whom I admire and respect as much.[19]

> He was a man, take him for all in all,
> I shall not look upon his like again.[20]

At the time when I arrived at the Galton Laboratory at UCL in October 1957 as a PhD student, Sylvia Lawler, a long-time associate of the Galton Laboratory had said: 'Anyone who managed to get a PhD in the Galton Laboratory had to have a streak of originality. There was no spoon-feeding. Penrose would take people in, shut them in a room, and let them get on with it.'[21] In my own experience, this was indeed so. During my second year at the Galton Laboratory, in April 1959, I told Lionel that I would not be staying for a third year. Lionel said that it was a pity that I had not written a PhD thesis. 'But, Professor,' I said, 'I am writing a thesis and you shall have it before I leave in October.' Lionel was not in the least taken aback; he wanted to know what the topic of the thesis was. I told him that the topic was a clinical and genetical study of the Pendred syndrome of deafness with goitre in collaboration with a physician across the road at University College Hospital, Dr W R Trotter. The mention

---

[19] Professor George Fraser wrote: 'His son, Professor Oliver Penrose, wrote of Lionel's tenure of the Galton chair at UCL between 1945 and 1965: "In particular, he was the first clinical geneticist, that is to say he advised patients who already had one mentally handicapped child what the chances were, if they had another child, that the new one would have the same affliction" (Penrose (1998): 12–13). Apart from these informal Saturday morning genetic counselling sessions at the Galton Laboratory, Lionel had long-standing contacts in the field of clinical genetics with colleagues at the neighbouring University College Hospital throughout his tenure of the Galton chair; towards the end of his tenure, in the 1960s, these contacts were formalized by a joint appointment in the field of clinical genetics.' Note on draft transcript, 28 February 2009; 2 March 2010. See also Professor Ursula Mittwoch on page 10.

[20] Thompson and Taylor (eds) (2006): Act 1, Scene 2, lines 186–7, page 182.

[21] Kevles (2007): 45; Robson (1996).

of Dr Trotter brings to mind the multi-faceted diversity of Lionel's intellectual interests, in that in 1952, in his role as a psychologist, he published a book called *On the Objective Study of Crowd Behaviour*, based to a substantial extent on the book *Instincts of the Herd in Peace and War* written in 1916 by Wilfred Trotter, the father of my collaborator. [22]

Of Lionel's many other talents, I shall not go into detail about his wooden self-replicating machines.[23] He was also an excellent chess player and delighted in the composition of chess problems. Lionel was a founder member and chairman of the Medical Association for Prevention of War. In a letter to the *Lancet*, adducing a scientific basis for the struggle for peace, he and six others wrote:

> Doctors have a social responsibility as well as a personal one to their patients; they have an ethical tradition and an international allegiance. War is a symptom of mental ill health. Its results include wounds and disease. Doctors are, therefore, properly concerned in preventing it. An 'epidemic of trauma' requires prophylactic treatment as much as any other epidemic.[24]

Lionel was brought up in a Quaker household and his aversion to war was already developed during his childhood, when he wrote a little poem:

> We are under one king
> Let our flag be peace.
> Let us take our swords off
> And let the bloodshed cease.[25]

During the First World War, he served in the Friends' ambulance train of the British Red Cross in France.[26]

---

[22] Penrose (1952); Trotter (1916).

[23] Clarke (1974): 246. A list of Professor Penrose's articles on his machines has been provided by Professor George Fraser and will be deposited along with other records of this meeting in archives and manuscripts, Wellcome Library, London, at GC/253.

[24] Doll *et al.* (1951).

[25] Professor George Fraser wrote: 'This poem was written by Lionel at the age of eight years (Smith (n.d., *c.* 1999): 6).' Note on draft transcript, 2 March 2010.

[26] Professor Lionel Penrose served in the Friends' no. 5 ambulance train of the British Red Cross in France in 1918. For further details, see Smith (n.d., *c.* 1999): 7.

An extract from the last section entitled 'positive eugenics' that he added to the fourth, new revised edition of his classic book, *The Biology of Mental Defect*, published in 1972, the year of his death, is revealing with regard to Lionel's views concerning the eugenics movement:

> It has been one intention throughout this book, to explore the broad problems of eugenics and to show that they cannot be solved until the mode of action of natural selection on the human race is much more fully understood than it is at present. It may be relatively easy to point to genes which appear altogether bad; for example, those which make their possessor victims of epiloia or of Huntington's chorea. Even these may not be unfavourable in all circumstances. Carriers of 'bad' genes may sometimes have compensating advantages....The position is quite different when we try to identify 'good' genes. The human types which are accepted by eugenists as desirable can be specified to some extent.... Nothing is known, however, about the actual genes which might form the basis of such qualities.[27]

I started my tribute to Lionel Penrose, pioneer of clinical genetics as well as of human genetics, with a quotation from Shakespeare and I will end with one from Osler:

> The great possession of any university is its great names. It is not the pride, pomp and circumstance of an institution which bring honour, nor its wealth, nor the number of its schools, nor the students that throng its halls, but the men who have trodden in its service the thorny road through toil, even through hate, to the serene abode of fame, climbing like stars to their appointed heights.[28]

I think both Shakespeare and Osler could have been thinking of Lionel when they wrote these words.

Bobrow: George, thank you very much indeed. And could I formally thank Professor Emery for helping with that presentation [Fraser dropped his papers during his contribution, Emery gathered them up and held them for Fraser to read].

---

[27] Penrose (1972): 293.

[28] Osler (1904): 10–11.

Professor Shirley Hodgson: I never actually worked with Lionel Penrose, but as he was my father, I knew him for a long time! I want to make a couple of brief points. An important influence on Lionel was what George Fraser has very eloquently mentioned, his Quaker upbringing. And I know JBS Haldane used to say that Lionel was deeply Quaker in all his beliefs except for the theology, and that thread went through much of his thinking. As Sue Povey said about the Galton Laboratory, he did try very hard to change its name, and he said from the beginning that he found it a continual embarrassment to be associated with the stigma of eugenics. According to the obituary from Harry Harris: 'The long delay [in changing the name of the chair] had been because of legal problems connected with the original wording of Galton's will.'[29] I know he very soon changed the name of the *Annals of Eugenics*. Its subtitle was 'a journal for the scientific study of racial problems' in 1926, and it was changed to 'a journal devoted to the genetic study of human populations' by Sir Ronald Aylmer Fisher in 1933, and Lionel changed the title to *Annals of Human Genetics*.[30] So he was constantly trying to change the emphasis and was very concerned about any association between what the laboratory was doing and any eugenic theories or associations. I would also endorse what Sue said about his enjoyment of being with people who had mental retardation. There was a lovely story he used to tell about how he was talking to a man with Down syndrome, which, of course, was called mongolism then, in the grounds of the hospital, and the chap with Down syndrome said: 'Oh, there's a mongol. There's another mongol. They must be digols'. I'll leave it there.

Harper: Could I just come in at this point and throw a question out to everybody, but particularly those people who had direct connections with the Galton: I'm always puzzled as to, with Penrose's particularly clinical approach and the fact that people came from all over the world to work at the Galton, many of whom would later be founders of medical genetics in their own countries, why did the Galton not become a centre of clinical genetics?[31] Because it didn't and from the 1970s on it looked at basic science, very distinguished basic science, but it wasn't a place that people looked to if they were developing clinical genetics. I'm not quite sure

---

[29] Harris (1973): 538.

[30] The journal was founded in 1926 by Karl Pearson, Galton professor of eugenics at UCL, who was succeeded by R A Fisher in 1933, followed by Lionel Penrose in 1945. See Clarke (1974); Li (2000); see also note 11.

[31] Professor Sue Povey wrote: 'The Galton Laboratory at UCL flourished, but the genetic clinic was not a major feature.' Note on draft transcript, 2 March 2009. See pages 16–17.

Figure 4: Gu Wen-xiang, Professor Bette Robson, Galton professor of human genetics, Dr David Hopkinson (Hoppy), director of the MRC human biochemical genetics unit, and Dr Joy Delhanty in the staffroom of the Galton Laboratory, 1981.

whether that came about because Harry Harris succeeded Lionel Penrose, or for other reasons, but I'd be interested to know if people can throw light on this.

Professor Marcus Pembrey: May I first give a second-hand comment about Lionel Penrose? I talked at length to Laurie Smith down in Colchester, a laboratory technician (1937–85) who used to help with the genetic clinics there and was appointed by Lionel in 1937 at the Eastern Counties Hospital (Royal Eastern Counties Institution, Colchester, Essex), as part of his MRC-funded work on mental retardation.[32] He used to be responsible for doing the urine tests for the new PKU test and I specifically asked him, I remember, and he said that Lionel did talk to the families before the war, where the nature of the inheritance was understood, PKU being one. Lionel then gave up the Eastern Counties activity and clinic to John Fraser Roberts.[33]

The other thing about why the Galton never flourished – if by the Galton you mean not the Kennedy-Galton (Northwick Park and St Mark's, London) but

---

[32] Professor George Fraser wrote: 'Lionel practised clinical genetics long before he took up the Galton chair. Thus he exercised the same functions in advising patients, as described by his son Oliver (Penrose (1998)), even during the Colchester survey of 1280 cases of mental defect, which he carried out between 1931 and 1938.' Note on draft transcript, 10 November 2009. See note 19.

[33] See Pembrey (1987); Polani (1987) and Fraser Roberts' biographical note on page 120.

the Galton Laboratory at UCL – was, I think, principally because John Fraser Roberts, Cedric Carter and people at the Institute of Child Health had already established a very active clinical genetic service in that region.[34] Also, the people who took over from Lionel, I think Harry Harris and Bette Robson (Figure 4), in particular, didn't welcome clinicians, if we put it bluntly. She said: 'We're proper doctors here', implying that the clinicians weren't. So I think that it was the style that led to it not being fought for, and there were adequate genetic services around.

Fraser: Lionel was a great man, and that's why he welcomed 'proper' (i.e. those with a PhD) doctors and 'improper' doctors (i.e. medical) to the Galton Laboratory. There is a great deal of biographical information available about his scientific activities both in human and in clinical genetics. His inaugural lecture as Galton professor in 1946 is very well known, and I had hoped that copies would be distributed to the participants at this seminar; I regret that this has not proved possible.[35]

Bobrow: I think that is right, but what we're trying to dig at is the history of the Galton Laboratory and its contribution to clinical genetics because there is a lot known about Lionel's own background.

Fraser: Then I'll restrict myself to one sentence from his inaugural lecture. He was talking about phenylketonuria, and he says: 'A sterilization programme to control phenylketonuria confined to the so-called Aryans would hardly have appealed to the recently overthrown government of Germany.' He didn't, in fact, know at that time what had happened in Germany. But this was a comment on the fact that phenylketonuria had not been found among Jews and Negroes.[36]

Bobrow: Extraordinary. I'm sure people will come back to get details from you.

Professor John Burn: Briefly on Gerald Corney (Figure 5) with whom I worked for several years, who is often forgotten.[37] He was a very gentle, very wonderful

---

[34] See pages 21 and 26–9.

[35] Penrose (1946). Professor George Fraser wrote: 'This lecture has been reprinted in the *Annals of Human Genetics* more than half a century after its first publication (Penrose (1998)).' Note on draft transcript, 28 February 2009. A copy will be deposited along with other records of this meeting in archives and manuscripts, Wellcome Library, London, at GC/253 and is available at www.ncbi.nlm.nih.gov/pubmed/9803263 (visited 8 March 2010).

[36] Professor George Fraser wrote: 'So much wisdom condensed into a single sentence – the infallible mark of a truly great man. And he wrote this at a time when very little was known of the excesses of the Nazi new order in Europe.' Note on draft transcript, 28 February 2009.

[37] See, for example, Burn and Corney (1984); see also note 14.

Figure 5: Dr Gerald Corney and Mrs Nona Parry-Jones, chief technician, Galton Laboratory, c. 1981.

doctor who did the most amazing pedigrees. He's the only person I ever saw who did a separate pedigree for each side of the family in his clinic. He had a considerable indirect influence: he was a great mentor to me for several years, and to both Peter Farndon and Robin Winter.[38] Many of us would go and sit and chat with him for hours. I think he had quite a lot of influence on our early development.

Professor Dian Donnai: Before ever I got into genetics, I was training in paediatrics in London, and I went to Northwick Park Hospital as it first opened in 1970. Lionel Penrose was at Harperbury Hospital for mental defectives, Shenley, Hertfordshire, at that time and I remember having him as the visiting consultant coming to our ward.[39] I've still got photographs of a patient with Rubinstein Taybi syndrome that he saw in 1970; and a patient with tuberous

---

[38] See, for example, Donnai (n.d., c. 2004). Professor Robin Winter developed the London dysmorphology database series with Professor Michael Baraitser, distributed from the mid-1980s and then published by OUP in 1990, followed closely by the neurogenetics database in 1991 (see www.lmdatabases.com/ (visited 19 January 2010); see also his biographical note on page 131.

[39] See Watt (2000).

sclerosis. My interest in genetics as a career was probably awakened by that contact. He used to bring a lot of people who seemed to be middle-European, who took fingerprints of everybody and looked in great detail at that, and there was Mike Ridler, the cytogeneticist, who was a lovely man and worked with Lionel.[40] But I have to say that I feel very privileged to have met Lionel at the end of his clinical career.

Bobrow: Let's move our way down the list. The next bullet is 'chromosomes and the need for clinical genetics'. I think when this was put down, there was a hope that there would be a small coterie of other cytogeneticists in the room who would speak to it.[41] I'm going to ask Joy Delhanty whether she has anything to say, and if not, I've got about two sentences and we might make up a lot of time on this.

Delhanty: I think I've already said my bit, so you can carry on with your two sentences.

Bobrow: Good enough. I think the point that I would want to make is that the development of clinical genetics at that time, in my view, was critically dependent on the discovery of the chromosomal abnormalities and methods for diagnosing them, because that, I think, gave a clinical need to see patients who were outside of the specialties of other major medical groupings. This was the territory that genetics then stamped out for itself: interpreting laboratory results, some of which were rather complex, and even pretty good paediatricians – they were all paediatric at that time – found it relatively hard to get to grips with the technology quickly. I think that theme has continued and expanded as the laboratory components of genetics, about which we are not to speak today, have become more and more complex.[42] The clinical geneticists have, in general, come to be an interface between the real world and laboratory results in a way that, so far, has required skilled interpretation. So each of the major centres at which clinical genetics developed and spread is associated with significant early contributors to the development of laboratory techniques. The two are, I think, absolutely interwoven. We've already heard about that for the Galton, but the

---

[40] Dr Michael Ridler was head of the Kennedy-Galton cytogenetics laboratory and co-chair of the Association of Clinical Cytogeneticists (ACC) with Dr Maggie Fitchett. See, for example, Mutton *et al.* (1996).

[41] For a history of cytogenetics and medical genetics, see Ferguson-Smith (2008).

[42] For the annotated and edited transcript of the 2001 meeting on genetic testing, see Christie and Tansey (eds) (2003).

same will, I think, be true for all of the other major centres. The labs and the clinics grew up together, tightly interwoven.

Mittwoch: Chromosome studies themselves were preceded by the discovery of sex chromatin bodies, and sex chromatin that is present in females and not in males. That gave rise to the discovery that patients with Klinefelter syndrome had the Barr body and therefore might have been females, and conversely patients with Turner syndrome lacked the Barr body. So, when chromosome techniques became available, they were the patients that were singled out for chromosome investigations.[43]

Harper: I suppose I'd better justify why I put this down in the first place. It's exactly what Martin Bobrow has said, that somehow there was interplay between the laboratory and the clinical side, and the two together were the main stimulus for clinical genetics developing from being something very small-scale to something much larger. I think it's worth remembering a quotation from Victor McKusick: 'Until the advent of clinical cytogenetics, our specialty was not yet born. We were dependent, like the fetus, but clinical cytogenetics gave us our own organ.' Actually, about six months ago, I quoted this in my book, and I thought I'd better find the source.[44] I couldn't find it anywhere, so I wrote to Victor saying, 'Where did you actually say this? It must be in one of your reviews,' knowing Victor, with his encyclopaedic memory, would definitely be able to find it. He wrote back to say that actually, he couldn't quite remember either, but I'm sure he did say it – maybe somebody can pinpoint that?

What Ursula has said reminded me of a fascinating experience I had a couple of years ago now, in recording an interview with Mike Bertram, who discovered the sex chromatin body with Murray Barr. He is now in his eighties, living in Toronto, Canada, and gave a vivid account of this discovery, as if it might have been yesterday. A blow-by-blow account, and you know, that was in 1948, published in 1949, but it's amazingly vivid.[45]

Dr Alan Johnston: If I could throw a little light on Peter's problem, because in 1960 I had just come back from spending over a year with Victor McKusick. I was full of enthusiasm, but there was no funding for clinical work by the

---

[43] Jacobs and Strong (1959); Ford *et al.* (1959a and b).

[44] Harper (2008): 282. See note 47.

[45] Barr and Bertram (1949). See Professor Peter Harper's interview with Ewart Bertram on 'Interviews with Pioneers of Human Cytogenetics', track two at www.genmedhist.info/Interviews/ (visited 26 January 2010).

authorities for me to become a clinical geneticist at UCH. I was told to go back to work as a senior medical registrar, which is what I did, and continued to do so until the end of my working life. There was definitely a financial element: laboratory work could be funded but not clinical work.

Pembrey: I would like to disagree a little with the complete emphasis on the cytogenetics and the origin of clinical genetics thereafter. I think there is another point to make regarding the province of clinical geneticists, not just in interpreting genetic-related tests, but pedigree analysis. And I think that the origins of the clinic at Great Ormond Street in 1946 by John Fraser Roberts and so on, first were based on pedigree analysis and then empirical risk figures.[46] He already had those for mental retardation at various levels from his work in Bristol, and with Cedric Carter this was then extended to congenital abnormalities of various sorts. So that was all occurring before the chromosomal abnormalities. So there was a need and pedigree analysis is still one of the things that is particularly the province of the clinical geneticist.

Bobrow: Point well taken. Should we move on to the Oxford unit and Alan Stevenson?

Dr Clare Davison: I'd like to tell you a bit about the background to this. Alan Stevenson was a professor of social and preventive medicine at the Queen's University Belfast in Northern Ireland, and I well remember as a student having these very peculiar lectures on genetics. We all thought them rather peculiar but we had to put up with it if we were to get that pass as part of our final exams. He was, at that time, very interested in looking at populations and population studies, and Northern Ireland in those days was a very captive area. All he had to do was to write to the clergy and say, 'I wish to carry out a study on such and such' and it would be announced at church and not a single member who went to church would refuse to take part in the study. A very good one was on myotonic dystrophy.[47] He then was appointed director of the MRC population

---

[46] Mrs Margaret Fraser Roberts wrote: 'It was 1946 when Sir Wilfred Sheldon, senior consultant paediatrician at the Hospital for Sick Children, Great Ormond Street, London, invited my husband to provide a clinic for genetic advice there.' Letter to Mrs Lois Reynolds, 18 March 2010. Professor Paul Polani's obituary of Dr John Fraser Roberts stated that 'in 1946, he established a genetic counselling clinic at the Hospital for Sick Children, Great Ormond Street, London (and later at the Children's Hospital, Bristol), the first ever organized clinic in Britain and Europe and the third in the world (the others being the Heredity Clinic, University of Michigan, and Dight Institute, University of Minnesota). In the same year he was appointed lecturer in medical genetics at the London School of Hygiene and Tropical Medicine.' Polani (1987): 309.

[47] Stevenson (1953).

genetics research unit in Oxford in 1958, and he was the only director there until his retirement in 1974. At the beginning there was a cytogenetics laboratory set up under Marco Fraccaro and the other medics at that time included George Fraser, of whom you've already heard, and John Edwards. Around that time the WHO did a worldwide study of congenital malformations under Alan's auspices. It was a great big, thick book, published in 1966.[48]

As well as the cytogenetics laboratory, there were two invaluable health visitors appointed then too. And, horror of horrors, there were Saturday morning clinics in the outlying hospitals and patients were also seen in local hospitals. The reason that they were Saturday morning clinics was that nobody else would have a clinic on a Saturday morning. I know a little bit more about what happened from 1962 because that's when a lot of us joined the unit. I could give you a list of some of the names of the people who worked at that unit from 1962 onwards, all those who were on the staff, whether they were MRC staff or whether they were visiting fellows from here, there, and everywhere – we'd lots of them – all set up departments, either clinical departments under the NHS or whatever the equivalent was in their own countries, or university departments. Some of them unfortunately are deceased: Charles Kerr, R S Wells, familiarly known as Charles, who was a dermatologist who came up to St John's here; I was there, Ian Shine, Martin Bobrow, Dick Lindenbaum, Derek Roberts, Denis Bartlett, Peter Pearson and Norman Nevin. My apologies to anyone I've forgotten.

I would like to mention some of the population work that Alan Stevenson was particularly interested in – he got his staff to do the work. His name may have gone on that paper at the end, but he made certain we did most of the work. One of the studies that I'd particularly like to mention was Charles Kerr's work on X-linked conditions in Oxford.[49] Charles was given the happy job of finding out in which families there were two or more males affected with something. He did his DPhil on that and, because he discovered that there were a whole lot of families with males with severe mental retardation, as it was known in those days, Charles went back to Sydney and he got Gillian Turner to do such a study,[50] whereas I did my MD on that in Oxford.

---

[48] Stevenson *et al.* (1966).

[49] Kerr (1968).

[50] Turner *et al.* (1970).

Another was Charles Wells's work as a dermatologist, using ichthyosis, one of the first papers with the delineation of the genetics, histology, clinical features etc. Charles enabled clinicians to assign the mode of inheritance based on the different clinical features of ichthyosis, and he was able to say, 'Ah yes, this is the mode of inheritance of the different types of ichthyosis'.[51]

I forgot to mention, and I think I'm right in saying, I can see that people were kind enough to say it was the first book – I don't know whether it was, Peter Harper is nodding in agreement – on genetic counselling. I looked at it the other day and I was horrified to find just how cheap it was. No wonder people were able to buy it and put it in their pockets and take it to clinics. It was published in 1970 by Alan and me.[52] We did another edition in 1976 and then, of course, there was Peter Harper's book, *Practical Genetic Counselling,* afterwards.[53] At that time, when I was in Oxford, we had population studies, Mendelian inheritance, empiric risks, and what I think is vital to all geneticists: intuition, because we frequently saw families where things occurred more than once, and what was very interesting was that a lot of those guesses, which were intuition, were subsequently proved correct by studies carried out by others. Alan did a lot for quite a lot of us, and the cytogenetics lab was set up with the main reason that we could persuade the doctors to let us see their patients. It was tit for tat: they referred patients to us and we saw the patients and did chromosome studies for them.

Bobrow: My recollection, subject to correction from people with better memories or bits of paper, is that the other name associated with that unit – the one you left out – is Jim Renwick, who I think never worked there. He signed up and was on staff for quite a long time, but didn't take up his appointment and took another job before he ever got around to starting. Still – interesting.

Two other comments about that unit where I spent my early years as well. One of the things that we had was a jolly good library and Alan was fanatical about getting every journal that one could conceivably get, saying that whatever needed to be read was there to be read on site. Secondly, he was extremely good at ensuring that everyone had a place in which they all had coffee together, and that kept the clinical staff of the doctors and the nurses and the lab staff all in touch with one another in a very convivial atmosphere, occasionally punctuated by a very aggressive atmosphere, but at least there was always an atmosphere that we

---

[51] Wells and Kerr (1965).

[52] Stevenson *et al.* (1970).

[53] Harper (1981, 2004).

were sharing. It was an interesting experience. That was an MRC unit built in the grounds of a hospital trying to interface with the 1970s NHS.[54] I suspect that it was a bit of a laboratory for both the MRC and the NHS. It wasn't unique, there were other places like that, but it was an interesting breeding ground for trying to work out how the interface between research, genetic service and clinical practice would work. There were health visitors working there from my earliest memory, which would be early 1960s, 1963, I think, or 1964.

Fraser: They were there in 1961.

Bobrow: They were there in 1961, so they weren't genetic counsellors, not only in name. I think – and we can get into this later – there were some differences in the way people were treated, but the concepts and the origins were there from a very early stage, there and elsewhere. The unit was closed by the MRC when Stevenson retired in 1974 and the people who were there, largely stayed. They largely stayed, with one or two exceptions. They stayed as either NHS or university employees.

Harper: Martin, you mentioned the library; you may be interested to know that about four years ago now I received a rather plaintive phone call from someone in the Oxford unit to say that there was no space to keep books any more so they were all going to be disposed of. This was actually the origin of our current Human Genetics Historical Library at Cardiff because I went over with my fairly capacious car and returned with around 300 books. It was only when I got back that I realized that many of them went right back into the 1950s and 1960s and had originated from the Stevenson time, and so I think the Oxford people and I were all very happy that this collection has now been preserved. That library's catalogues are available on the internet and so you can see exactly what it consisted of.[55]

Might I be allowed a question for Clare Davison? You're absolutely right, Clare, yours was the first book on genetic counselling from the UK, although we mustn't forget Sheldon Reed's from the US.[56] But I'm always intrigued to know about the origins of the chapter in it on Bayesian risks, which I'd been

---

[54] Harper and Pierce (2010).

[55] The Human Genetics Historical Library was established in 2002 at the Institute of Medical Genetics, Cardiff University, supported by funding from the Wellcome Trust, and is now part of Cardiff University's special collections and had over 3000 volumes in 2010. See www.genmedhist.info/HumanHistLib/ (visited 29 April 2009).

[56] Reed (1955).

Figure 6: Dr John Fraser Roberts at the time of his election as a fellow of the Royal Society, 1964.

brought up with in Baltimore from Tony Murphy.[57] Was this something that Alan Stevenson or you derived, so to speak, *de novo*, or did you get it mainly from reading Tony Murphy's papers on Bayesian risk estimates?

Davison: Alan did most of that. I remember Alan doing the risks and he asked round the rest of us whether we agreed or not. We also had a statistician on the staff at that time and he was sourced to vet it. Actually, there are two different ways of working out the risks: there's the way that you did it and the way that Alan did it. And if you actually do the same thing twice, one way and then the other, the risks come out the same. I'm not sure that it was all that easy for people to understand. Most of the doctors that I met said, 'We skip that out; we go to the other more interesting chapters.' Martin Bobrow will probably put me right on that.

Bobrow: No, I can't actually. I certainly know that Alan was doing Bayesian calculations and that he loved them. He showed me how to do them, and I have to say I didn't love them that much. But I don't know whether he worked it out for himself or read it in one of his books.

---

[57] See, for example, Murphy and Mutalik (1969); Bayes (1763).

Davison: I think before he came to Queen's University, Belfast, he worked for some time under the auspices of R A Fisher.[58] I may be wrong on that, but he certainly worked for someone in this.

Bobrow: There's no question that he was highly numerate; he was a public health epidemiologist by training. He didn't have trouble with statistics. What I don't know is whether he took the idea of Bayesian risk calculations from Murphy or not, but he did them, undoubtedly, in vast quantities. Let's move on to John Fraser Roberts (Figure 6), Cedric Carter (Figure 7) and the Institute of Child Health.

Pembrey: I'm delighted to have Margaret Fraser Roberts next to me and I'm sure she can fill in some of the details. I knew John Fraser Roberts well because I spent a decade, many days a week, going through editions of his book, a first edition of which I have here, the *Introduction to Medical Genetics,* 1940.[59] This in fact is quite interesting – the archivists might be interested: John's book, in which he pencilled in all the changes he was going to make for the 1959 edition, which was the second edition. There was a lot in this book that actually spelled out pedigree analysis in particular and the tools of the clinical geneticist. But it wasn't until the 1959 edition that there was an actual section on the clinic, talking to patients and so on. But I think it's well known, and Margaret can fill in the details here, that while he was working at the London School of Hygiene and Tropical Medicine he was asked to start a genetic clinic at Great Ormond Street in 1946 and he did so.[60] Fairly soon afterwards, certainly by the late 1940s, he was joined by Cedric Carter, who was a research fellow there, focusing on congenital malformations. That clinic continued and then in 1957 John Fraser Roberts moved with his team to the Institute of Child Health at UCL to become the first director of the new MRC unit for clinical genetics

---

[58] See, for example, Stevenson (1954). Dr Clare Davison wrote: 'I think it might have been Bradford Hill (1897–1991) rather than R A Fisher (1890–1962), but I am not certain.' Note on draft transcript, 13 February 2009. No mention was made in the Royal Society biographical note of either.

[59] Fraser Roberts (1940). See his biographical note on page 120.

[60] Great Ormond Street Hospital had a tradition in this field. Mr Nick Baldwin, GOS archivist, wrote: 'Sir Archibald Garrod (1857–1936), generally considered to be the founder of modern genetics research, was a senior physician at Great Ormond Street. We have 29 volumes of his patient casenotes from 1899–1913. Among other things, these contain the first recorded case of alkaptonuria in 1904, and there may well be other "firsts". The volumes under 100 years old still have restricted access to external researchers on data protection grounds, but the others are open to bona fide researchers.' E-mail to Mrs Lois Reynolds, 10 March 2010. See also notes 4 and 46.

there.[61] Margaret will, I think, enlarge on something that's interested me about clinical genetics: the origin of the sort of weekly meeting, or the way in which the information was gathered in advance of going to the clinic. And also the generation of separate genetic notes, which is something we've retained. I know in my time we were struggling to maintain that separateness, because we saw not just one person but the whole family. The notes go back, I think, to the practice at Great Ormond Street and the Institute of Child Health in the late 1940s and 1950s.

The research into the genetic counselling, fairly directly, was first instigated by John Fraser Roberts and then Cedric Carter, establishing the consecutive case series from Great Ormond Street, where they looked at recurrent risks by diagnosis. That gradually became the basis on which they gave genetic counselling to families where there wasn't a clear inheritance pattern. John moved to Guy's Hospital, London, in 1964, and the thing that I particularly remember about his period there, apart from collaborating on the book and getting to know him very well, was his evidence-based medicine, as it would now be called. He was funded to do follow-up studies by the MRC. And so there were a series of follow-up studies, the results from which were published, in which he found out what the families did after coming for genetic counselling and relating it to what they were told.[62] Now Cedric carried on or cooperated in those studies, and in fact, interestingly, it was a result of those follow-up investigations about what people actually did in relation to what they were told at the clinic, that I remember and quite definitely changed Cedric's way of counselling.[63] I remember him telling me, and I remember discussing with Martin Bobrow and Michael Baraitser, that he became concerned that the way the counselling had been done in the past was putting too many people off having babies and so he tempered it with a qualifying phrase: 'We regard this as a very low risk' or something like that. This actually came from the follow-up.

John came from Stoke Park Hospital, Bristol, so he maintained his outlying clinic there and also he took over from Lionel Penrose at Colchester. So certainly John was doing at least those two regular outlying clinics from London and I think there were others for a short time in Plymouth and Taunton, and one or two others. As far as I know, but I don't have this very firmly, Cedric Carter didn't

---

[61] Pembrey (1987). See also page 21 and note 46.

[62] See, for example, Carter *et al.* (1971).

[63] See, for example, Dennis *et al.* (1976).

Figure 7: L to R: Mrs Kathleen Evans, Dr Nick Dennis and
Professor Cedric Carter, 1981.

do many outlying clinics. I may be wrong, but what I can glean from talking
to people is that Cedric was basically based at the Institute of Child Health
and used to do clinics at Great Ormond Street and the National Hospital for
Nervous Diseases, Queen Square, London.

I've been interested in the fact that John would tell me quite often that he only
qualified in medicine in order to study humans, to do medical genetics, and the
day he qualified he hung up his stethoscope. I can remember him saying that
repeatedly. He therefore made it quite clear that he wasn't going to be diagnosing
patients and that he gathered together good evidence of the diagnosis and said
that clinical genetics always had to be practised in a place where you could get
expert diagnostic skills. I wondered whether that might have been the origin of
this tradition, at least at Great Ormond Street, of the weekly meeting at which
everything was gathered together. But I do think that at least in Great Ormond
Street, the origin lies much more with Margaret Fraser Roberts and the way she
wrote out a format for the clinic for visiting doctors to follow. I leave that for
Margaret to say.

With regard to the assistance of the genetic counsellors or others, fairly early on
in the 1950s, Kathleen (Kath) Evans, who was an almoner at Great Ormond

Street Hospital[64] when Cedric was setting up things there with John Fraser Roberts, volunteered to go into the genetics side. So, that was from the social work area, and she gradually took on a role preparing information before the clinic and then talking to the families before they went to see Cedric, and quite often, because of Cedric's rather perfunctory style, talking to them afterwards as well. I think Kath teamed up with someone named Anne, from Victor McKusick's outfit, who used to come over from Johns Hopkins.

Cedric carried on this style of clinical, empirical risk figures, meticulous family studies with the assistance of family visiting personnel, health visitors or people with social work backgrounds at the clinic at Great Ormond Street. But he did a lot to encourage the formal recognition of clinical genetics as a specialty, and I remember as clear as day, the day clinical genetics was on the Department of Health's list of clinical specialties. I think it was 1980. And he came through from his room into the library, waving it, saying: 'We're in! We're in!' (or something along those lines). So he was very keen to get clinical genetics recognized. And of course he was instrumental, along with Dr Sarah Bundey and others, in setting up the Clinical Genetics Society.[65]

Mrs Margaret Fraser Roberts: Marcus Pembrey very kindly referred to a bit of help that I used to produce before my husband's clinics for genetic advice. He only needed brief details of a case. I used to produce notes for him so he knew what sort of conditions were coming up, in case he wanted to look anything up beforehand. At Great Ormond Street Hospital and at Guy's Hospital in particular, he started having visiting doctors from other countries. Very often there was a language problem, so I just expanded my bare-bones notes a bit so that there was a little précis about each family or patient who was going to be seen, and that seems to have gone down quite well, I think.

Dr Nick Dennis: I joined Cedric Carter as what I think was probably the first registrar in clinical genetics in the country in 1972 and stayed until 1976. It was an MRC unit, as has been said, headed by Cedric. As well as Kath Evans, there were people like Jean Heath, Becky Coffey and later, Joan Warren. They were primarily employed for research purposes, because the main function of this unit was research. They all went out and gathered family data on whatever abnormality was being studied. So, the clinic was run rather as a sideline and it's true that Kath Evans did attend those clinics, but coming in as registrar,

---

[64] An almoner is a medical social worker. See Baraclough (1996); Bell (1961); see also pages 30 and 71.

[65] See note 7; see also Burn (1983).

my main job was to look at all the referral letters that came in and do the background reading so I knew what cases were coming to the clinic. But I don't ever remember sitting down with Cedric before the clinic and having any sort of meeting to discuss what was going to be said. In three-and-a-half years I think I probably sat down and spoke to a family less than six times, so my training was very much as a fly on the wall. The reading round was very helpful, but I had no real hands-on experience. There was a weekly clinic, which I think was on Thursday afternoons, and there would be a list of eight families. So they were given about 20 minutes each and Kath Evans and I would sit in and occasionally there would be another visitor sitting in. Cedric was fairly brief, but very focused. I only remember perhaps three or four times that Kath Evans had to rush out and comfort somebody in the corridor because they were visibly upset. So although her presence there was comforting, she didn't have a formal role, as far as I can remember, either seeing people before or after the clinic.

Other points worth mentioning: in this unit of perhaps a dozen or 15 people, there was also Joan Slack, who had a rather separate existence running a lipid research laboratory carrying out population studies and she was collaborating with Northwick Park at the time.[66] Everybody got together for tea and coffee at ten o'clock in the morning and three o'clock in the afternoon for ten or 15 minutes and that was a three-line whip; you didn't miss that. I found it very helpful. Everybody would down tools for ten or 15 minutes, and go to the library to talk. Michael Baraitser turned up about half way through my time there. He'd come over from South Africa where he was a neurologist and he'd developed a neurogenetics interest and I think he took off some of the Huntington's disease families to see at Queen Square. The main sort of case mix was the congenital malformations, where Cedric was very much on his home territory, giving an empirical recurrence risk. Then what we later started calling 'query syndrome' cases. I remember those as being pretty much a snap judgement on Cedric's part without a very prolonged clinical examination. Then came the chromosomal abnormalities and then Huntington's disease.[67] Anne Child was there before me, so perhaps I wasn't the first registrar. I think Cedric regarded her as his registrar as well, and I was succeeded by Anita Harding and then John Burn.

Burn: I was the last clinical scientific officer – I think that was my official title – before Carter retired. And it's worth pointing out the distinction

---

[66] See for example, Slack *et al.* (1977); see also Reynolds and Tansey (eds) (2006): 28–30.

[67] See discussion on page 76.

between genetic consultation and counselling. Cedric wasn't a genetic counsellor; he did genetic consultation. I always tell people in Newcastle that he only gave them 12 minutes, whether they needed it or not. But the point was that he didn't see it as his job to explain in detail. He would give his opinion and then write to their doctors and it wasn't in any sense that he wasn't being a good doctor, he just didn't have very much time. His perception was that he should give his expert opinion and then pass it back to the patients' own doctors. So I think that was, in a sense, a distinction from what we now perceive to be our task, explaining it in words of one syllable to the doctor and to the patients.

Bobrow: That's a very interesting distinction. I wonder what was going on elsewhere. My training was to be told to do the Swindon clinic one Saturday morning a month and I got a bunch of notes, and I read what I could, I got on my motorbike – I had a wonderful red Honda four-cylinder bike at the time – and I went to Swindon once a month and saw people. Bear in mind that I had never studied genetics or anything much else – I was a junior registrar in surgery before I made this lateral move. Then I rode back to Oxford and tried to sort out the mess afterwards. Alan Stevenson was always available if there was something I wanted to talk about, but usually I would start with Clare Davison and progress to him if the need arose, and he would spend as much time as it took to talk through individual cases. But there was no question of structured training at all in that sense. I would say that we were definitely groping our way in terms of counselling. We saw our job as trying to ensure that the patient understood the diagnosis so that they could explain it to the doctor.

Mrs Lauren Kerzin-Storrar: I want to add something to Marcus Pembrey's comments about the follow-up studies that were done to see what reproductive choices patients made after attending genetic counselling clinics. I think that the discrepancy between what the risk figures might predict about behaviour and what families actually chose to do led Cedric Carter to engage a psychiatrist called John Pearn. Nick (Dennis), was that when you were at Great Ormond Street? Is it right that John Pearn was a psychiatrist?

Dennis: I think he was a paediatrician – he did the work on spinal muscular atrophy (SMA), didn't he?[68]

---

[68] See, for example, Pearn (1973b); Pearn *et al.* (1978).

Kerzin-Storrar: I didn't know that, but he wrote a seminal paper on psychological mechanisms influencing interpretation of risk that is regarded as salient today.[69] I understood that that study initially came out of an interest in understanding why patients weren't doing what the doctors thought they were telling them to do and therefore was the first step, perhaps, towards developing a non-directive ethos in genetic counselling practice.

Pembrey: Yes, I can confirm that John Pearn was involved. In fact, I remember taking him off to the Long Lane antique market (Bermondsey, London) very early one morning, because that was one of his interests as well. And I think that's probably right, the paper came out of that discussion.

Harper: A point about Cedric Carter and eugenics: Cedric was, I think it's fair to say, one of very few and possibly the only practising clinical geneticist who had quite strong eugenic inclinations, at least in recent years. People may disagree, but I think it's fair to say that most of us felt distinctly uneasy about that. But Cedric didn't bring eugenics into his regular practice; he had these views, and I think most of us realized he did have them, and most of us, I think, agreed to differ on it. Is that a fair perception?

Dennis: Well, I remember Cedric saying once that there was an article he was going to write for *The Times*, but he'd have to have been retired before he wrote it – I'm not sure it ever appeared. But Cedric regarded himself as a 'positive eugenicist'; he was interested in increasing the intelligence of the population, so he thought all of the more intelligent people should have a lot of children. He had seven.[70] He once asked me how many children I had, and I said: 'Two.' He looked a bit thoughtful and said: 'Hmm, I think you should have another.'

Bobrow: That's nice: he thought well of you.

Dennis: He certainly never tried to discourage people in the clinic from having children. He was very non-directive, I always felt. Another thing I would like to say about him personally is that although he wasn't a very talkative character – he was always rather terse – his door was always open and you never felt any sort of bar to going in and sitting down with him and he would always give you as much of his time as you needed.

---

[69] Pearn (1973a).

[70] Dr Nick Dennis wrote: 'He thought that if contraception was readily available, the less intelligent would voluntarily curtail their own fertility. He did not believe in any form of coercion or overt discouragement.' Note on draft transcript, 15 February 2009.

Figure 8: Professor Paul Polani in Cameron House, Guy's Hospital, initial home of the paediatric research unit, 1975.

Fraser: In this connection there is a paper written by Cedric Carter and a gentleman called Kenneth Hutton, who taught me chemistry at Winchester College in the late 1940s.[71] They were both scholars at Winchester College and I

[71] Professor George Fraser wrote: 'The paper was read at a meeting of the Eugenics Society on 18 November 1952, at a time when Cedric Carter was secretary. His influence was acknowledged in the publication as follows: "A large number of other friends and colleagues have assisted…and I should like to thank them all very much, but most especially our secretary, Dr Cedric Carter, who has provided the driving force and encouragement necessary to get this paper produced." (Hutton (1953): 215). As stated, both Cedric Carter and Kenneth Hutton had been scholars of Winchester College, a group characterized in the paper as "an extremely intelligent section of the community (page 215)…exceedingly few of them not being in the top 1 per cent of the population, with respect to Cattell IQ (above 160 or 170)." (Hutton (1953): 207). The importance of the inquiry was stated as follows: "It appears probable that the intelligence of the nation is declining but there is little or no direct evidence in this country of how many children people of high intelligence have….Here in ex-pupils of Winchester College I have a sample of high intelligence and I have found out how many children they have." (Hutton (1953): 205). The number of the other scholars' children was slightly lower than the number of children born to males in the general population. Nevertheless, the paper ended on an upbeat note: "The most encouraging sign is that among the scholars – who cover a 20-year range – there is evidence of a steady increase in family size so that it may be expected that the 1921–25 group, (i.e. those born in 1908–13) will have not merely more children than the average of the population but even just enough for replacement." (Hutton (1953): 215).' Selection from a note on draft transcript, 19 September 2009; e-mail to Mrs Lois Reynolds, 2 March 2010. The complete version will be deposited along with other records of the meeting in archives and manuscripts, Wellcome Library, London, at GC/253.

don't know where this paper was published, it may be in the Winchester College magazine. They were very enthusiastic about counting the number of children the scholars of Winchester had, and obviously they were keen that they should have as many as possible. As you say, Cedric Carter had seven; Kenneth Hutton had four. I'm not sure that the other scholars kept up with them.[72]

Bobrow: Should we move down our agenda to Guy's? I'll ask Caroline Berry to begin.

Dr Caroline Berry: Paul Polani was the leading spirit at Guy's and he was a real polymath.[73] If it hadn't been for the Second World War, I think he'd have been a neurologist and stayed in Italy. As it was, he joined the British merchant navy, he was interned in the Isle of Man and spent the rest of the war as a surgeon in the Blitz at the Evelina Children's Hospital.[74] Then, in 1948 he was for some reason persuaded to become a physician and he became a quite a leading paediatrician at Guy's. But he was interested in the origins of children's conditions. He used to work with the children's cardiologist Maurice Campbell,[75] and they noted, I suspect it was Polani who noted, that the girls with Turner syndrome had coarctation (narrowing of the aorta), but coarctation was a condition that was more common in males. Therefore they questioned whether Turner's was some kind of pseudosexual anomaly. Then, of course, came the finding of the Barr bodies, which confirmed the suspicion, and then chromosome analysis.[76] So, he led straight from paediatrics into genetics because it seemed to show the origins of the disorders.

Polani's curiosity and enthusiasms were in a way rewarded because in 1960 what was then called the Spastics Society (Scope since 1994) endowed the Prince Philip chair of paediatric research at Guy's. They must have put a lot of money into the paediatric research unit, which was housed in a warehouse at the back of Guy's. Polani became its director and this gave him full rein to

---

[72] See also Carter (1956, 1966).

[73] See Harper (2007); see also Christie and Tansey (eds) (2003): 6–12, 15–20, 26, 50, 100; Figure 8.

[74] The Evelina Hospital, London, founded by Baron Ferdinand de Rothschild in 1869 in memory of his wife who died in childbirth, was amalgamated with the children's unit at Guy's Hospital in 1947 and moved into a new building at St Thomas' Hospital in 2005. See www.guysandstthomas.nhs.uk/services/managednetworks/childrens/evelina/about/history.aspx (visited 13 January 2010).

[75] Silverman (2003).

[76] Polani *et al.* (1954, 1956); Ford *et al.* (1959b).

pursue his research interests. There were chromosome labs, others where he had Philip Benson doing biochemistry and Mary Seller doing embryology. He had a brilliant library so that people didn't waste time going to the Guy's library. Then, when John Fraser Roberts left Great Ormond Street, Polani inveigled him to join so that there was also a clinical aspect.

Polani was an excellent paediatrician and he continued to see children mostly – he was an early syndromologist and he was very skilled. If I'm allowed one anecdote: I remember being rather shocked when we were seeing a family where there were two children with heart disease, one of whom had died, and he said to the mother: 'Would you remove your clothes?' She duly did, and she was covered in lentigines, freckle-like spots. Polani had noticed that she was a little deaf, and this was an early family with what we used to call Leopard syndrome (now called multiple lentigines syndrome).[77]

However, what happened then was that the prenatal diagnosis scene burst upon us. At Lewisham they were doing amniocentesis for Rhesus, so the Guy's cyto lab got the samples and found it could karyotype amniotic fluid cells, and therefore prenatal diagnosis became possible.[78] Not long afterwards David Brock found alpha-fetoprotein (AFP). I can remember sitting at the kitchen table reading the exciting paper and thinking: 'Wow, if this is something, this is big news.'[79] Then, just down the road we had Stuart Campbell using ultrasound to see fetuses *in utero*.[80] This was all burgeoning and becoming more clinical than Polani wanted, so he took on board, first of all, Jack Donald Singer, and then I joined the team and, quite unlike Nick Dennis, I was funded to see patients from the word go. Although Polani continued to see the difficult paediatric cases in the clinic, a big clinic developed. We did a lot of prenatal work, a kind of overflow from the John Fraser Roberts clinic. Again, all this was on soft research money; I was a research assistant. There was no NHS money and there

---

[77] Lassonde *et al.* (1970); Blieden *et al.* (1981); Shamsadini *et al.* (1999).

[78] Santesson *et al.* (1969); Christie and Tansey (eds) (2003): 16–19. See also Medical Research Council, Working Party on Amniocentesis (1978). Dr Caroline Berry wrote: 'Rhesus sensitization was being monitored by amniocentesis by the late 1960s, so I expect this was when it started at Lewisham. Sadly the obstetrician responsible, Mary Anderson, died a few years ago. The first chromosomal diagnosis on amniotic fluid done at Guy's was in 1972, and they would have needed at least a year to develop the technique.' E-mail to Mrs Lois Reynolds, 1 March 2010.

[79] Brock and Sutcliffe (1972).

[80] Tansey and Christie (eds) (2000): 30, 43–4, 54–5; see also Campbell and Kohorn (1968).

was a constant struggle to get clinical recognition. So, that was how it all started at Guy's.[81]

Bobrow: Who else remembers Guy's at that time? I went to Guy's in 1983 – that's so recent it's not history!

Harper: Might I ask Caroline Berry a question? At what point did the regional development of clinical genetics services start to happen? Because I always see Guy's, perhaps along with Manchester, as having pioneered the concept of covering the whole region, rather than just thinking in terms of one central focus.

Berry: I think it must have been about 1978. I remember going with Paul Polani to the NHS regional offices in Croydon and talking to them there to persuade them. It was interesting because an evening television programme at that time ran a feature on prenatal diagnosis, which was very exciting. I was there and was interviewed on that programme and the chap we saw at the Croydon office had seen it and obviously realized that this was not going to go away. So, something probably came out of that in the late 1970s.

Pembrey: Actually, the origins of the regional side at Guy's came from a research study that was aiming to identify all trisomies in the south-east region.[82] In fact, the nurse appointed to do the family visits and take blood was Dottie Garrett and that began in the 1960s. So, it was on the back of that research project that all the regional links were established and the flow of clinical samples started. I think that is probably why it was seen as very early in terms of being organized regionally. That was in the 1960s and 1970s.

Davison: Surely Oxford was one of the first ones for this, because in 1974 the NHS took over the MRC unit, and at the same time in Cambridge, under the East Anglian regional health authority, genetics and cytogenetics laboratories were established, as were clinics, in three of the district hospitals.[83]

Bobrow: Well, they are two separate things. When were regional clinics being done? It was organized on a hub-and-spoke basis and Oxford was certainly

---

[81] Polani *et al.* (1979).

[82] The paediatric research unit was established at Guy's Hospital, London, in 1960, the Supraregional Laboratory for Tissue Enzymes in 1973 and the South-east Thames regional genetics centre in the unit in 1976 (Polani *et al.* (1979)). See Hamerton *et al.* (1965); see also Polani *et al.* (1960).

[83] Dr Clare Davison wrote: 'Peterborough, Norwich and Ipswich district hosptials.' Note on draft transcript, 27 February 2010.

Figure 9: Sir Cyril Clarke.

running regional clinics regularly in 1964/5. There was a whole ring of peripheral hospitals that people visited regularly. When it was taken over by the NHS it was all done on MRC money – only moderately appropriately – and when the unit closed in 1974 it was formally taken over as an NHS commitment.

Davison: The East Anglian regional genetics service was set up in 1974.

Bobrow: Should we move down our list to Cyril Clarke and the Liverpool Institute?

Professor Sir David Weatherall: It's always difficult to talk about Cyril Clarke (Figure 9). It's like travelling in Ireland, when you ask the way, they say: 'Well, you can't start from here.' I think many of you know Cyril's background: he came to Liverpool in the early 1950s as a straightforward consultant physician, building up a private practice and with a major interest in asthma. There was no hint of things to come. But from his early childhood he had had a profound interest in butterflies and, later, genetics. He was breeding butterflies furiously by the early 1950s and got fascinated with mimicry. Purely by chance, through an advertisement for samples of butterflies, he met Phillip Sheppard from Oxford, who was one of E B Ford's protégés and they became very close friends.

Cyril told me, and I think he told several people, that it was while walking on the Broads that Phillip said to him: 'Doesn't all this genetics have any applications to medicine?' They threw this thought around and Phillip said: 'The only thing that can be easily studied at the moment is blood group genetics.' This was around the time that Ian Aird from the Hammersmith Hospital had produced his preliminary data on the relationship between blood group A and gastric cancer.[84] So, they started work in the field of blood group associations, working with duodenal ulcer and secretor status. Along the way they became interested in Rhesus, and put a young chap called Ronnie Finn to work on a rather mundane kind of problem for his thesis: to try to confirm previous work suggesting that ABO incompatibility might protect women against Rhesus sensitization. I do not need to expand on that story any further because you dealt with how it led to the prevention of Rhesus disease very thoroughly in a previous Witness Seminar.[85]

When I was Cyril Clarke's houseman in 1956, medical genetics was just starting to zoom, lots of excitement, everybody rushing around the hospital spitting into tubes and smelling their urine, and lots of very bright young people coming along for research projects. After I left for my national service, Cyril established a connection with the Moore Clinic at Johns Hopkins Hospital in Baltimore with Victor McKusick.[86] David Price Evans went from Liverpool as the first of the Moore Clinic fellows and did some rather beautiful work on pharmacogenetics.[87] I followed David to the clinic as the next fellow when I came out of the army. Meanwhile, at Liverpool, a number of other registrars were coming along with bright ideas and new projects, and because of the

---

[84] In 1935, Professor R A Fisher and Dr G L Taylor at the Galton Laboratory started a genetic study of blood groups, with a Rockefeller Foundation grant and MRC support, later constituted by the MRC as the blood group research unit in 1946 at the Lister Institute, directed by Dr R R Race (Thomson (1975): 150). See Aird *et al.* (1954); Clarke *et al.* (1956); see also Weatherall *et al.* (1985).

[85] See Zallen *et al.* (eds) (2004): 28–30; see also Finn *et al.* (1961).

[86] The Moore Clinic at Johns Hopkins Hospital was endowed by Dr J Earle Moore as a chronic disease clinic in 1952 and became Dr Victor McKusick's clinic for the new division of medical genetics in the department of medicine in 1957. See McKusick (2001). For an interview with Professor Victor McKusick and his wife Dr Anne McKusick by Andrea Maestrejuan for the oral history for the medical genetics project at UCLA in 2001, see also www.socgen.ucla.edu/hgp/mckusick_interview.html (visited 17 February 2010).

[87] Price Evans (1962).

connections with E B Ford and the Nuffield Foundation, it was decided to support Cyril in building the Nuffield Institute of Medical Genetics.[88] So, by the mid-1960s when I came back from the US, the building was almost ready and we were able to have a complete floor to develop haemoglobin genetics. David Price Evans was there and there was a cytogenetics group led by Stanley Walker on the top floor. Above this, on the roof, there was a large butterfly breeding facility. My own team was greatly strengthened by the arrival of John Clegg from the Laboratory of Molecular Biology in Cambridge and together we were able to evolve a strong team which carried out some of the early work on haemoglobin synthesis in thalassaemia, the pattern of fetal haemoglobin production during early development, and a programme which laid the way for the first description at the DNA level of a deletion as the cause of a common genetic disease.[89]

There was also some excellent work going on in the department of medicine at that time with a major focus on genetics. For example, immunogenetics was particularly strong through John Woodrow's work on the human leukocyte antigen (HLA), diabetes and rheumatoid disease.[90] If you look at the history of the development of HLA-related disease genetics, Woodrow's early work is often overlooked, but it was absolutely seminal in developing a solid basis for the HLA relationship with type-1 diabetes.[91]

Some of you in this room will ask, as many others have, why the Nuffield Institute was never established as a more formal clinical genetics centre.[92] I think the answer is clear when seen in the light of Liverpool medicine's view on specialization during the postwar period. I remember Cyril Clarke's predecessor, Lord Cohen, asking me what I was doing with my career and when I told him I wanted to do haematology he looked at me as though I had just crawled out from under a stone. 'A specialist is a person who knows everything about his field except its relative importance', he added. Most of the clinical basis for genetic work in Liverpool during Cyril's time came from clinics in related fields. Our haematology clinic where I saw patients with inherited blood diseases

---

[88] See Anon (1963); see also Zallen *et al.* (eds) (2004): 46; Harper (1985).

[89] Clegg *et al.* (1968); Ottolenghi *et al.* (1974); see also Weatherall (2001).

[90] Cudworth and Woodrow (1974); Woodrow (1975, 1985, 1988); see also Zallen *et al.* (eds) (2004): 23–8, 30–6, 41–5.

[91] See, for example, Gale (2001).

[92] See earlier discussion on the Galton Laboratory on pages 15 and 17.

was one. The genetics of rheumatic disease came from John Woodrow's big clinical practice in rheumatology and all of Cyril's work on tylosis and cancer of the oesophagus was based on work in the gastroenterology department at Broadgreen Hospital, Liverpool.[93] I do not think that Cyril ever believed that genetics should become a separate specialty from the rest of medicine. Rather, and I think this was his enormous strength, he felt that genetics had relevance to every aspect of medicine and it should stay like that. It was only when we had all left Liverpool that clinical genetics *per se* developed. But, apart from his own wonderful work on Rhesus disease,[94] Cyril's great contribution was to inspire groups of extremely talented young people – some of whom are in this room by the way – about the importance of genetics in medicine, but without trying to formalize the association in any way.

Bobrow: He actually went on doing that for a long time. I never worked there, but I met him at the Royal College of Physicians (RCP) and other places. He was inspirational.

Professor Alan Emery: Coming from the north of England, I have to reply. My boss, Sir Robert Platt,[95] and David's boss, Cyril Clarke (Figure 9), were very good friends. It was Robert Platt who encouraged me to do genetics. But I think we have to remember what it was like then. In a weak moment before coming to this meeting, I found my inaugural address when I went to Edinburgh in 1968, and I got a quotation from the *Lancet* in 1966 that says: 'Medical geneticists able to speak with equal authority among both pure geneticists and consultant physicians are, not unexpectedly, as rare as astronomers who can navigate a liner.'[96] One tends to forget where you're coming from: if you're in inner London and you've been with all these very famous people, of course the scene is set, but if you were outside it was very different.

I came into medical genetics much later in life because I'd been in the army, and then taught for several years before studying medicine. So when I qualified and completed junior hospital jobs, I was 33, nearly 34. What are you going to do then? My boss said: 'Go to see Cyril Clarke and talk about medical genetics.' That was an eye opener because there were all these very famous people doing

---

[93] Howel-Evans *et al.* (1958).

[94] Clarke and Sheppard (1965).

[95] Platt (1963, 1972).

[96] Emery (1968a): 12.

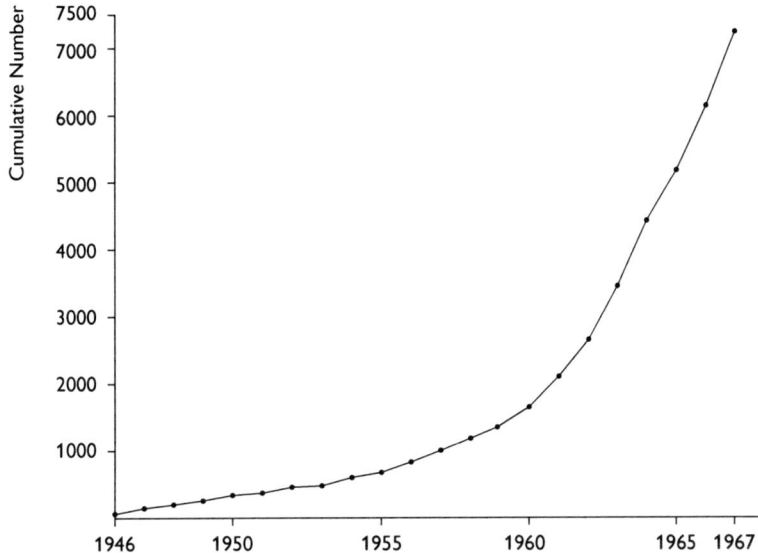

Figure 10: Total number of published papers on human genetics, 1946–67.

various aspects of genetics, which encouraged me. But it was a long time before the experts got hold of this. And it was people like Cyril and Robert Platt who, I think, did it more than anyone else.

Cyril Clarke convened a meeting at the Royal College of Physicians of London when he was president (1972–77), for people who wanted to do medical genetics. There are perhaps several people in this room who were at the meeting. He took a current textbook on medical genetics and waved it in the air: 'All you need to do is read this, but get your Membership first.'[97] I think in those days we were trying to get people to be card-carrying physicians or surgeons and then to become geneticists. Of course, the picture has changed now, but that was how it was in the early 1960s (Figure 10). I owe a great deal to Cyril, even though he got me carting those damned great boxes about from Liverpool to Manchester to collect *Biston betularia* (the peppered moth).[98] I never found any of the black ones, because I lived in the suburbs and there wasn't any atmospheric pollution. Apart from that, we owe a lot to Cyril Clarke and to Robert Platt, who got it going at a clinical level.

[97] Professor Alan Emery wrote: 'This was Emery's *Elements of Medical Genetics* (Emery (1968b)), now in its fortieth year in 2008 and 13th edition), which was awarded the BMA best medical student textbook prize in 2008.' Note on draft transcript, 16 January 2009.

[98] See, for example, Kettlewell (1956).

Donnai: In looking through the archives of medical genetics in Manchester, I came across a letter dated 22 November 1958 and signed by Robert Platt, regarding the formation of a group for the study of human genetics (Figure 11). He talked about it being his intention to do this, and he said that as he was shortly to give up the chair of medicine, he hoped to have more time to devote to this kind of work, and that his appointment as chairman of the recently constituted MRC committee on human genetics should enable him to keep in touch. There were quite a lot of people at the Paterson Institute attached to the Christie Hospital involved in early cytogenetics. I've also found the medical board records when Alan Emery asked to set up a Wednesday clinic in clinical genetics in 1964. In fact, when I joined Rodney Harris we still had that clinic in the Manchester Royal Infirmary. The group for the study of human genetics was formed and was very healthy until at least 1969/70 and had very eminent people speaking at it. Alan Emery fostered the group for a while and then Rodney Harris took it over in 1968 with enthusiasm. I found quite a lot of Rodney's records of all of the people whom he had put a great deal of effort into inviting to speak there. So, I think Manchester has a history almost as long as Liverpool's, spearheaded by Alan and then Rodney.

Professor Rodney Harris: When I told Cyril Clarke that I might be leaving Liverpool to go to Manchester, he said to me: 'You'd be a mug if you go to Manchester.'

Bobrow: Was he right? No. Let's roll seamlessly into Manchester: how did it all happen in Manchester? Who set what up?

Emery: I have an anecdote about Rodney Harris I would like to relate. When he came to take over my job in Manchester he found a bottle of magnesium trisilicate (for indigestion) in my drawer and said: 'My God, is this a difficult job then?' But it was Cyril Clarke and Robert Platt who got it going. Robert published quite a bit on the genetics of hypertension and he had this multifunctional debate in the literature of the time, 'Platt vs. Pickering', as to whether the distribution of hypertension was multifactorial or unifactorial.[99] That encouraged a lot of people. But I have to emphasize, going to Manchester in the 1960s after I had spent some period at Johns Hopkins, the funny thing was that there wasn't a lot of interest. I could talk at a meeting and half of the consultants would be asleep! But they have no excuse now for doing that, whereas they probably did then.

---

[99] Swales (ed.) (1985); see also Ness *et al.* (eds) (2002): 37, 66.

Department of Medicine,
The Royal Infirmary,
Manchester, 13.

22nd November 1958

Formation of a group for the study of Human Genetics.

Dear

It has long been my intention to form in Manchester a group of people interested in the study of human genetics. The first object of the group would be to bring together people working in different branches of medicine who have a common interest in human genetics, for instance, there are to my knowledge people working in clinical medicine, general practice, psychiatry, statistics, serology, histology, paediatrics, pathology, rheumatism, obstetrics, etc., who are interested in this subject. By meeting together periodically, and hearing short informal accounts of work going on in various fields, we might be able to pool information to our mutual benefit, and we might in some instances be able to share resources, for instance by calling upon research funds for the appointment of a social worker or laboratory technician who might work for more than one member of the group.

As I will shortly be giving up the Chair of Medicine I hope to have more time to devote to this kind of work, and my appointment as Chairman of the recently constituted M.R.C. committee on Human Genetics, should enable me to keep in touch with work going on in various parts of the country.

I am calling a preliminary meeting to be held in the Dayroom of M 1 Unit, Manchester Royal Infirmary, on Saturday, December 13th, at 11.30 a.m., and hope you will be able to attend. We will then discuss the desirability of forming a group on these lines, and those present at the first meeting may be able to suggest the names of others who might be interested. Please reply on the form below.

Yours sincerely,

Robert Platt.

SUGGESTED AGENDA for the first meeting: members may like to come prepared to discuss very briefly possible developments in one or more of these subjects:-

Human Cytology, including nuclear sex, in relation to genetics.

Study of abortions.

Twin Studies.

Hereditary disease, with especial reference to linkage and the detection of
                                                            heterozygotes.
Selective advantage of heterozygotes.

The genetics of normal human traits and of common diseases.

Infertility clinics in relation to genetics.

I intend to come to the meeting.

I cannot come to the meeting.

I cannot come to the meeting on December 13th, but would like notices of future meetings.

(Cross out words not required).

Figure 11. Invitation to discuss the formation of a group for the study of human genetics, 1958.

We appointed John Timson to study chromosomes.[100] I believe this was the first centre outside of London that was a proper medical genetics centre. We had two technicians doing biochemistry, measuring creatine kinase, which I developed when I was in the US, then the Boehringer Chemical Company took it over commercially, otherwise I wouldn't be here now, I'd be in the Bahamas! This became our way of doing it, and counselling became terribly important. I went to see Cedric Carter, who was very helpful to patients. Victor McKusick was, as I say, in the US, and talked a lot to patients and that came over very strongly. It wasn't just a question of diagnosing the condition, though this had its own specialties, and, of course, needing material from other laboratory sites in order to help. The main thing was to converse with the patient. People, even senior medical students in Oxford, still ask me now what to say if the patients are a Catholic family and you've got to talk about prenatal diagnosis. These are the questions that we started to look at in Manchester, as we had to deal with that problem with a lot of Catholics there. There were also a lot of orthodox Jewish people who also couldn't consider many things like this and I learnt an awful lot from that maelstrom, that multifaceted society in Manchester, which maybe I would have got in London too.

I went up to Edinburgh in 1968 and that was when the university got together with the health service, the Southeast Health Board in Scotland and the MRC, to fund a proper department doing nothing but human genetics – medical genetics. Someone's mentioned David Brock's AFP test[101] that took off then and John Scrimgeour's fetoscopy.[102] We were all going into this new world of what it meant to patients with genetic diseases. Looking at the inaugural lecture I wrote in 1968 – I found by chance a couple of days ago – which shows how much we depended on people in London.[103] Nearly all of my references had to do with Paul Polani, John Hamerton and people like this, who did so much work to give clinical genetics a scientific basis. And, of course, we shouldn't forget Harry Harris, because he got inborn errors off the ground, which I thought was terribly important.[104]

Harper: Could I just reinforce the point that David Weatherall made about Cyril Clarke? I think it has been very fundamental and long-lasting. Cyril

---

[100] See, for example, Harris *et al.* (1969).

[101] See, for example, Brock (1979).

[102] Scrimgeour (1978).

[103] Emery (1968a).

[104] See, for example, Harris (1970).

genuinely saw genetics as part of everybody's job in medicine, and yet, to be honest, I think most of the people in Liverpool and elsewhere – most of the other physicians – didn't know what he was talking about, and didn't care too much. In a way, he was absolutely right, but he was 40 years too early. I'd go one stage further and say that of the various people that Cyril was instrumental in training, very often in conjunction with Victor McKusick – most of us went into medical genetics as a specialty – retaining a greater or lesser connection with general medicine. David himself, in Oxford, has been the only one who has been able to put Cyril Clarke's vision into practice. I don't honestly think that would have been possible anywhere other than in Oxford, simply because you needed to have an array of talent across the board, which just about every other centre didn't have. It is only now that Cyril's original vision is becoming generally accepted. That's how I feel anyway.

Harris: I'm quite embarrassed about this bit of discussion because I ended up in Manchester under completely false pretences. I'd never been trained in genetics and I was offered a job in Manchester to do tissue typing in support of the renal transplant unit. It was at that point that Cyril Clarke said that I'd be a mug to go to Manchester and that if I'd stayed in Liverpool I could have been a physician. However I did go, and attached to my tissue typing job was, I think, one Saturday morning clinic on genetic counselling, which I didn't understand at all. I did that, but it grew and grew and I can only say that my connection with medical genetics was entirely accidental, but very entertaining. Incidentally, you (Martin Bobrow) may have had a four-cylinder Honda – I had a two-cylinder horizontally opposed BMW – a much better bike!

Bobrow: I agree to that.

Donnai: I think the department in Manchester grew a great deal under Rodney Harris's influence. Rodney got to know everybody in the region and he worked enormously hard to build up a credible clinical service with NHS funding, recognizing that anything that was on short-term funding would be a disaster. We also began the regional genetic register service in Manchester and Lauren Kerzin-Storrar was one of the early appointees to develop that service.[105] Rodney also fought very hard to establish one of the first properly and permanently funded NHS senior registrar posts. There were two others in the country: one was in Cardiff, to which Ian Young was appointed, and the other was at the Kennedy-Galton Centre and Robin Winter (Figure 12) was appointed to that.

---

[105] Manchester's genetic register service was established in 1980. See also Kerzin-Storrar (1996).

Figure 12: Professor Robin Winter, c. 1992/3.

I'm pleased to say that Rodney appointed me to the third proper training post in 1978, although at that time there was no formal curriculum or anything. You still got sets of notes that you learnt from and you also learnt from the patients. I think the diagnostic laboratory and molecular laboratory services were built upon various initiatives that may be mentioned a bit later today.[106] Rodney was always very enthusiastic about the regional service and from very early on there were clinics set up in peripheral centres and because Manchester is in the south of the northwest region we went up as far as Lancaster until we strayed on to the Newcastle territory.

Burn: Derek Roberts, of course, who we heard mentioned earlier, moved to Newcastle in, I think, 1969, as a lecturer in child health, and became the first lecturer in human genetics. Very early, with Surinder Papiha in particular, he started to expand into medical genetic laboratory work and could see that as a way of funding the service.[107] Val Davison, now lab head in Birmingham, started her career as a cytogeneticist in the department. Derek was very much a research-orientated population geneticist, having previously been a geographer, but he developed a clinical service with the help of Rosemary Boon, the wife of Tom

---

[106] See, for example, Harris *et al.* (1989).

[107] Papiha and Roberts (1972); Lanchbury *et al.* (1990).

Boon, then a leading physician.[108] They took on two part-time health visitors, Jean le Gassicke and Dorothy Gibson, to set up peripheral clinics. Derek continued to do genetic counselling until his retirement.[109] I gradually eased his fingers off the handlebars because he hadn't done any medical training at all and, in fact, he was a little nervous of the medico–legal issues, so he tended to be hesitant about making a clear diagnostic statement. But, nevertheless, he was very influential in getting us going, getting the clinical service working and properly resourced, and getting us a small terraced house in which to be based. I joined him as a medical student back in 1972 to do a degree in genetics and then went to Victor McKusick in 1974 and decided to announce that I would be a clinical geneticist. About that time the university offered George Fraser the job – and I thanked George many years later because George chose not to turn up; he decided to stay in Ottowa instead. They were so annoyed at this that they decided to wait until I qualified in 1984, and gave me the job. I thank you again, George.

Bobrow: There is a big institute of neurology in Newcastle, which was doing an awful lot of stuff to do with genetic disease. Was there any contact, until recently, I mean? – I know there is now, but in those days were these just two separate entities?

Burn: Yes, I think they were. John Walton, now Lord Walton of Detchant, obviously got heavily into Duchenne dystrophy and did that himself, establishing a local reputation for muscle genetics which persists to this day.[110] Derek Roberts undoubtedly had some involvement, but he was mostly seeing referrals from paediatrics. One small interesting anecdote was that I inadvertently distressed Rosemary Boon enormously when I returned from London, because she had always assumed that the diagnosis she was given by the paediatrician was the one she should counsel to. I told her that we didn't believe anything anyone said when they referred patients to our clinic and we almost always changed the diagnosis. I think I hastened her early retirement by telling her that.

The only legal case I have ever faced originated in the team's willingness to accept diagnoses; I told a lady that she had pseudo-achondroplasia, a type of short-limb dwarfism, and that it was autosomal dominant, which was why she had an affected child. It turned out that many years earlier she had been told

---

[108] Boon and Roberts (1970).

[109] See, for example, Roberts (1968).

[110] See, for example, Bobrow *et al.* (1988); Christie and Tansey (eds) (2003): 10, 33–4, 41. For Lord Walton's biographical note, see page 130.

by Derek Roberts on the basis of paediatric diagnosis, that she had Morquio syndrome, a rare type of dwarfism with serious consequences, and it wouldn't recur. When she did have a recurrence, she went on to sue the hospital. But, since I'd taken over as head of the service, I was the one who was named. I decided that God was punishing me. For the record, the court ruled the case to be inadmissible, because the doctors had done the best they could at the time.

Bobrow: Let us move on to bodies that were involved in the field, the first of which was the Royal College of Physicians genetics committee and the Special Advisory Committee (SAC). There are several people here who know all about that.

Johnston: I'm not quite sure how it all started, but the Clinical Genetics Society produced quite a number of reports in the 1970s.[111] One of the things discussed was the question of accreditation of the specialty. The Joint Committee for Higher Medical Training (JCHMT, replaced by the Joint Royal Colleges of Physicians Training Board in 2007) was pretty well ruling the roost in terms of training, in allocating training posts. And they had set up these various committees. I remember very well having Paul Polani at one of the meetings discussing these reports and I was unwise enough to say that I was expecting to see Sir John Crofton the following week in connection with general medicine. Paul Polani immediately picked this up and said: 'Can you go and talk to him about clinical genetics?' So, I ended up having coffee after breakfast with Sir John in the Station Hotel in Aberdeen and that was the basis of the SACs. I talked to Sir John and he was very supportive until we came to the question of having our own SAC. He explained that the whole JCHMT had taken a policy decision to restrict the number of SACs and to reduce the number of those involved as far as possible. From that point on, we were stuck with the problem of supervision of accreditation and that wasn't solved during my professional lifetime. Things have changed quite a lot now, I understand, but it presented us with all sorts of problems in trying to supervise training schemes of senior registrars in clinical genetics.[112] And just for the record, the second senior registrar for clinical genetics in Scotland came to Aberdeen.

Emery: The first approved specialist registrar in genetics was in Edinburgh.

Harris: (getting the microphone) Thank you very much indeed. I thought you'd cut me off because I said my bike was better than yours.

---

[111] Dr Alan Johnston wrote: 'Initially at Cyril Clarke's request.' Note on draft transcript, 18 February 2009.

[112] For details of current training arrangements, see www.jrcptb.org.uk/Specialty/Pages/ClinicalGenetics.aspx (visited 20 January 2010).

Bobrow: No, I admire you for having a better bike. I couldn't afford one; MRC rates, you know.

Harris: It was quite an expensive bike, especially when I broke it. Alan Johnston was a marvellous guy to work with on the RCP committee on clinical genetics. He was very good indeed. What we did was to push the registrar of the RCP, David Pyke. Well, that was the SAC, and there wasn't an SAC, as you've explained, until we had finished and moved on. What we did was to turn up regularly at the JCHMT, which I think must have been the most boring part of anybody's work – it was not the most imaginative. However, it was good because it would get senior registrars coming into the specialty and there was an increasingly good training structure. They didn't play around and hope for the best.[113]

I think the most important single event at that time was when it was agreed at the King's Fund forum that there should be a single representative body for clinical genetics, including clinicians and laboratory people, and nurses and others of that ilk. That is actually published and the reference to that is on one of the bits of paper on your seats there.[114] What followed after the Clinical Genetics Society was the British Society for Human Genetics (BSHG). That was wonderful because it got all of the different specialties on equal terms, exactly as the King's Fund had envisaged. Another thing that exercised us a great deal at that time was the question of getting clinicians and laboratory people to work closely together. It's hard to believe now, but there was quite a lot of struggle going on over who controlled laboratories. The idea of it all being 'under one roof' took hold. Now, it was not literally under one roof, but it was spiritually under one roof.[115]

---

[113] Professor Rodney Harris wrote: 'Other initiatives at that time included successful lobbying of the NHS chief scientist to establish genetic associates as integral members of the NHS regional clinical genetic teams. The need for a British association for genetic medicine was stressed to include genetic associates as well as genetic nurses, medics, scientists and laity in a single representative body. See Harris (1988). This heralded the British Society for Human Genetics.' Note on draft transcript, 23 February 2009. See also Appendix 1, page 84.

[114] Harris (1988).

[115] Professor Rodney Harris wrote: 'Close collaboration between cytogenetic laboratories and clinicians had not always been the case so we encouraged the organization of clinical and laboratory genetic services "under one roof" to promote better management of patients and their families and training and research. The artificial separation of medical genetics from medicine, paediatrics, obstetrics, pathology and other fields has also been diminished and the committee on clinical genetics surveyed the education and training of clinical geneticists. See Royal College of Physicians, Clinical Genetics Committee, Working Party (1990).' Note on draft transcript, 23 February 2009. See Appendix 1, page 84.

Another thing that was extremely good and rewarding was the dialogue that we had with the ministers about genetic associates. Derek Roberts says that at that time he told genetic associates that they didn't have a pay structure except what they had to find themselves, like nursing. That was the beginning of getting genetic associates put on the same salary scale as laboratory scientists. That was quite a pleasing thing. Another development that was also a result of a lot of discussions with ministers was the establishment of 'special medical development', which was NHS funding for DNA labs in the health service.[116] As a result of that, three centres – Cardiff, the Institute of Child Health, London, and Manchester – were set up officially with NHS funding and that took off.[117]

I think the next thing is to go from things that were referred to as a medical-genetics sort of specialty to an idea that a lot of us have had, and I think I felt very strongly about coming from a department of medicine, being a physician by training, that medical students needed to be taught genetics. The RCP established a survey of all medical schools.[118] Each dean was written to and asked what genetic teaching was going on in their school. I can remember at the time, somebody at the college – I can't remember who it was – saying: 'Oh, we've never been involved in teaching.' But we did it. And it was very successful; whether it actually improved teaching, I don't know. People will know that better than I do. Alan Johnston himself took on postgraduate medical genetic teaching as well.[119]

The business of people other than geneticists doing genetics is, of course, a very real and fundamentally important activity. So we set up a National Confidential Enquiry into counselling for genetic disorders by non-geneticists. The most important observation that came out of it was that hospital records in paediatrics, obstetrics and other specialties were so bad we couldn't find out.[120]

---

[116] Department of Health and Social Security (1987).

[117] Professor Rodney Harris wrote: 'During the 1980s and 1990s my colleagues and I shared specific concerns about medical genetics in the UK and the EC generally. Although DNA methods were becoming important in clinical genetics, adequate resources were not yet available in the NHS and an application was therefore made for NHS Special Medical Development funding to pilot NHS DNA laboratories in Manchester, Cardiff and London, see Harris *et al.* (1989).' Note on draft transcript, 23 February 2009.

[118] Royal College of Physicians, Clinical Genetics Committee, Working Party (1990).

[119] See, for example, Johnston (1978, 1979); Anon (1983); Griffiths (2006).

[120] Professor Rodney Harris wrote: 'Subsequently in a series of NHS funded confidential enquiries we audited the quality of genetic counselling and prenatal diagnosis, the most striking finding of which was the poor clinical records of genetic management in obstetrics and other specialities that limited audit. The subsequent wide dissemination of these reports was aimed to draw the attention of professionals to these deficiencies. See Harris *et al.* (1999).' Note on draft transcript, 23 February 2009.

However, certain very good things did follow from that: one was the question of prenatal screening, and one of the things we looked at was screening for haemoglobin disorders, and one of the great things that followed on from that was NHS screening programmes for haemoglobin disorders[121] (haemoglobinopathies, common inherited blood conditions affecting people whose origins are from Africa, the Caribbean, the Middle East, Asia and the Mediterranean, but are also found in the northern European population).

The final thing was getting out into Europe. We established a concerted action on genetic services in Europe, which contacted all 31 European countries, to find out how they did it. Did they recognize genetics as a specialty? What was going on? That produced a report, and I'm told by European colleagues it is now widely used.[122]

Johnston: I mentioned earlier that we had produced a series of reports on training and staffing by the Clinical Genetics Society. This culminated in a big meeting at the college in May 1976, where there were representatives of the various health departments and the professors of medical genetics, and so on.[123] It validated our reports of the Clinical Genetics Society to substantiate this steady involvement of other departments. So we particularly wanted to be sure that everyone who was at that meeting accepted all that was being asked of us.

---

[121] The NHS Sickle Cell and Thalassaemia Screening Programme was set up in England in 2001 following Government commitment in the *NHS Plan* (Department of Health (2000): 109) to the world's first linked antenatal and newborn screening programme. See http://sct.screening.nhs.uk/; and www.library.nhs.uk/ SCREENING/ViewResource.aspx?resID=268964 (both visited 21 January 2010). See also: Thein and Weatherall (1987); Lane *et al.* (2001).

[122] Professor Rodney Harris wrote: 'Medical genetics was recognized as a specialty in the UK in 1980 thanks to the efforts of Dr C O (Cedric) Carter. I endeavoured to extend this recognition and was able, with European colleagues, to obtain EC funding for a concerted action on genetic services in Europe, a comparative study of medical genetics in 31 countries. (Harris (ed.) (1997)). This set up a network of clinical geneticists to stimulate recognition of medical genetics in all European countries and to investigate the structure, workloads and quality of genetic services and the different responses made in each country to the new opportunities and challenges of clinical genetics.' Note on draft transcript, 23 February 2009. See also Appendix 1, page 85; CGS Working Party (1983); Figure 14.

[123] Speakers at the half-day conference on 13 May 1976 were: Sir Cyril Clarke, Professors C O Carter, A E H Emery, J H Hutchison (chair), M A Ferguson-Smith, Drs A W Johnston, D F Roberts and H Soltan. Dr Alan Johnston donated his papers relating to the preparation and organization of the 1976 meeting and other documents on the training and function of medical geneticists, both nationally and in Scotland (1973–97). These will be deposited along with other records from the meeting in archives and manuscripts, Wellcome Library, London, at GC/253.

Harris: Ian Lister Cheese is in the audience somewhere and I have to say that our relations with the Department of Health were superb, partly at least because he was such a good intermediary.

Bobrow: Now, the next item on our list is the Clinical Genetics Society that then morphed into the British Society for Human Genetics. We're going to start with Michael Laurence, please.

Professor Michael Laurence: The CGS started off as a clinical genetics society at an informal *ad hoc* meeting in Edinburgh. We all came up to Edinburgh, sat round a table and, in fact, we decided to have a secretary. Cedric Carter was going to be the leading secretary and he had some guidelines, I think, of some society based in each of the main centres. To some extent, we were having to cater for the surrounding districts as part of the NHS, not as part of the MRC or some other body. And this was enthusiastically received and supported. I think if we hadn't done that we would still be in a mess today. The NHS supported us quite fully on that.

Bobrow: Who was at that first meeting?

Laurence: Rodney Harris (Figure 14), Cedric Carter (Figure 7), Alan Johnston, myself (and I think there may have been Norman Nevin from Northern Ireland). It wasn't a large meeting.

Bobrow: And Sarah Bundey? Yes: Peter's just whispered 'Sarah Bundey'. That's very helpful, thank you. Does anyone have anything to say about the CGS in its early days, because I then want to move onto the BSHG and the transition?

Pembrey: I have an anecdote about accreditation.[124] I think it's worth recording here that the CGS produced a number of early documents.[125] One was the *Provision of Services for Prenatal Diagnosis of Fetal Abnormality in the UK*; and then *The Provision of Regional Genetic Services* in 1978; *Provision of Regional Genetic Services in the UK* in 1982 (Figure 13); and then in 1983 *Report of the*

---

[124] Professor Marcus Pembrey wrote: 'When accreditation was being considered in 1976/7, Sir Cyril Clarke was president of the Royal College of Physicians and he advised me to get accredited in general medicine and in medical genetics rather than seek dispensation as an established medical geneticist. This went smoothly and I was accredited in both subjects in 1977. In 1978 I received a letter from the college saying something like: "We are planning accreditation in medical genetics and have discovered, to our surprise, that you are already accredited. It would help us in our deliberations, if you could you give us details of your training."' E-mail to Mrs Lois Reynolds, 4 March 2010.

[125] Dr Alan Johnston wrote: 'Chaired by different people.' Note on draft transcript, 18 February 2009.

**THE PROVISION OF REGIONAL GENETIC
SERVICES IN THE UNITED KINGDOM**

Supplements to the Bulletin of the Eugenics Society

No. 1 (1975)   Quantity and Quality of the British Population. 50p.

No. 2 (1977)   Assortative Mating in Man: Husband/Wife Correlations in Physical Characteristics. £1.50.

No. 3 (1978)   The Provision of Services for the Prenatal Diagnosis of Fetal Abnormality in the United Kingdom. £1.50.

No. 4 (1982)   The Provision of Regional Genetic Services in the United Kingdom. £1.80.

Report of the Clinical Genetics Society Working Party on
Regional Genetic Services

by

J. S. Fitzsimmons (Chairman), M. Baraitser, B. C. C. Davison,
M. A. Ferguson-Smith, N. C. Nevin and M. E. Pembrey.

Figure 13: Report of the Clinical Genetics Society's Working Party, 1982.

Working Party on the Role and Training of Clinical Geneticists.[126] Another reason for flagging it up is that these were reported in supplements of the *Bulletin of the Eugenics Society* and I think that, going back to the earlier discussion, this is often a little bit of a surprise for people who didn't understand that the Eugenics Society at that time was a natural assistant to getting these things published and so on. And that I think people sometimes overinterpret the fact that these early documents were published as a supplement – it was just a way of getting them published. There were a lot of people in both organizations.

Harper: Just to follow up on what Marcus Pembrey has said: I was on the council of the CGS at that time, and I think I'm right in saying that the main reason for the Eugenics Society publishing things was that the CGS had no money at all in its early years, and the Eugenics Society had lots of money and didn't know what to do with it; it was a very pragmatic decision. But those reports, and likewise the series of reports from the RCP Clinical Genetics Committee, were very influential right across the world, especially in continental Europe, not just in Britain. And, I think it is important a complete collection is made, which is something we are trying to do with the Human Genetics Historical Library. So, if any of you have got copies of those old reports, again, check the index on the library website to see if it is in the library, because these are just the kind of publications that get lost sight of, in comparison with conventional books.[127]

---

[126] Clinical Genetics Society, Working Party (1983).

[127] See note 55.

Professor Heather Skirton: I'd like to make a comment before we move on to the BSHG, about the huge support the CGS gave to genetic nurses in the early days.[128] We did have our own very small society but the CGS gave us a lot of moral support and also practical support. We used to hold our own scientific meetings under the auspices of the CGS, I can't remember whether it was the first day of CGS, or the last. And through our clinical genetics colleagues we got a huge amount of support that helped us to move forward as well.

Bobrow: My recollection – and I was never a big player – is that it was a very well-meaning and well-intentioned organization. From quite early on, personally I was not greatly drawn to an organization that made quite as much distinction between clinical and laboratory and other professions as the CGS was bound to do by its structure. I guess a lot of people felt that way, which is where the drive came from to slowly move forward to the BSHG. I don't know who's going to talk about BSHG in its early days, but John Burn seems a good person to start.

Burn: Yes, I'd like to speak about this because it's probably the only true historical contribution I can make.

Bobrow: You're too young, that's the trouble.

Burn: When I was made secretary of the Clinical Genetics Society by Alan Johnston and a cartel of senior colleagues in 1989, I don't remember being given the chance to vote on the decision, but I did say that I needed Peter Farndon to be my colleague as treasurer, so we split the job in two. I immediately thought – and Peter and I can't quite remember how this came about, but we agreed – that we needed a joint conference, and I think it was my idea to go for creating a joint society. I remember suggesting it to Peter at a ceilidh at one of the meetings, I think in Cardiff, and he said: 'Well, that's a good idea if you can pull it off', or something to that effect. I set about trying to persuade the other societies. The CGS and the nurses and counsellors were all fine, but the Association of Clinical Cytogeneticists (ACC) proved to be a harder nut to crack, but eventually they came on board.[129] In fact, the plan was to merge completely, but the ACC reneged and said they wanted to stand apart because they were a friendly society. So we created the federation model, and in fact, to give him credit, my colleague in Newcastle, John Wolstenholme, representing the ACC in discussions, pointed out that linguistically it should not be the

---

[128] See Appendix 2, pages 87–8.

[129] Professor Sir John Burn wrote: 'The cytogeneticists were always less likely to agree. They were fiercely independent.' E-mail to Mrs Lois Reynolds, 8 March 2010.

Figure 14: Dr Hilary and Professor Rodney Harris, 2006.

British Society *of* Human Genetics, but the British Society *for* Human Genetics. So we stand apart from the Americans and the Europeans on that point. We started in 1996, which is when I finished my tenure as secretary.

Donnai: I think we should, in this regard, pay credit to Andrew Read, who was the first chairman of the British Society for Human Genetics.[130] I remember having a talk with Martin Bobrow, probably in a bar somewhere, and as we were talking about the formation of this new British Society for Human Genetics, Martin said, in his very decisive way: 'The only person that can be the first chairman is Andrew Read.' And I think he was absolutely right, because Andrew

---

[130] Professor Dian Donnai wrote: 'Andrew Read joined the Manchester department as lecturer in 1978. He made major contributions to early studies on prevention of neural tube defects by folic acid and multivitamin supplementation and to the development of prenatal diagnostic testing. He was instrumental in setting up the research and diagnostic molecular genetics laboratory in Manchester from 1982 and his group discovered *PAX3* mutations in Waardenburg syndrome (Tassabehji *et al.* (1992)). He established collaborations with clinicians and scientists internationally and is widely recognized as a brilliant teacher. He was a founder member and then chairman of the Clinical Molecular Genetics Society and the founder chairman of BSHG. *Human Molecular Genetics* (Strachan and Read (1996)), now in its 4th edition in 2010, has become a major text in its field and *New Clinical Genetics* (Read and Donnai (2007)) is in its 2nd edition in 2010.' Note on draft transcript, 5 March 2010.

commanded the respect of all the constituent societies and everyone, and I think that was an inspired choice. So much so that he insisted on being called 'chairman' and not 'president', and didn't have any regalia to go with it either.

Bobrow: Yup, and you're right: it is respect that does it.

Pembrey: I think another influence, about that time or just before, was Pat Jacobs, who was president of the Clinical Genetics Society.[131] She wasn't a physician and that helped as well. I think that was important.

Bobrow: Right, we can move on from that. So the next thing on the list is labelled 'Department of Health' and I think what we're trying to get at is the relationship that several people have referred to already, between the department, whose face was often Ian Lister Cheese, but none of us understood what was going on behind you, Ian, and this developing society, or societies, all of whom were somewhat jockeying for position. And some of us, I guess, were slightly naive about how things work in the real world as well. We thought it would be interesting to hear how this looked from the other side of the fence. To give notice to others, I'd quite like, under this heading, because it doesn't come in anywhere else, to touch a little on screening, in a general sense, for genetic disease. I know that Bernadette Modell covered a lot of that in the other meeting.[132] I'd just like to get a few words on the screening programmes because they came out of the same stable. Ian, how did it look from the Department of Health? Why did anyone let us get away with setting up an operation of this nature?

Dr Ian Lister Cheese: Having been invited as a non-specialist and non-expert – and here by false pretences, as so many other geneticists are, of course – you might wish to know that people like me came to the Department of Health at a mature age, often with a good deal of clinical experience (I'd been in general practice in Wantage in Oxfordshire, a tutor and trainer in general practice, and had a small role in management in the cottage hospitals around Oxfordshire). One of the things that was very vivid – I think Hilary Harris will know what I mean – is that when you have responsibility for something like 1600 or 2000 patients in a market town and village community, you don't attend 1600 patients, you attend 100 or 200 families. You know their dispositions, their diatheses, as the Victorian physicians used to say. In a sense I've always had that feeling, as it were, for the genetics of the population – the disposition of the population.

---

[131] See, for example, Jacobs *et al.* (1959); Jacobs and Strong (1959).

[132] Christie and Tansey (eds) (2003): 48–9; see, for example, Modell B (1997); Modell M *et al.* (1998).

So I came into the Department not as an expert, but as a generalist when Donald Acheson became Chief Medical Officer (CMO, 1983–91).[133] After the usual sort of induction that new civil servants have to undertake in rather run-down country houses to be taught how not to embarrass their ministers and the like, he asked me to take over the policy job that covered children in hospital, the newborn and genetics, and to become secretary of the Standing Medical Advisory Committee. So genetics was obviously a very small part of what I was asked to do.

I read into the subject, as one must; there wasn't very much within the Department. The Standing Medical Advisory Committee had, in 1972, produced a very small booklet on genetics.[134] It was a description of Mendelian genetics with a few pictures of chromosomes, and a little paragraph about genetic counselling.[135] I thought at first, since it was produced in 1972, that Cyril Clarke, who had become president of his college at that time, might have produced it. Now I know there was no possibility whatever of him having written that document. The point was this: there was very little by way of a policy statement, very little indeed. And it didn't matter, because the work of the professional bodies, the RCP, the Clinical Genetics Society, and of individuals – many of them – had led to a series of authoritative documents which laid down clearly what a clinical genetics service should comprise. They set out its purpose, its function, and all the elements that make up an effective clinical genetics service. They described its educational and research roles and they had investigated the ethical aspects of genetics. There was a sense in which one didn't have to do anything. The ground had been laid.

This was very important in another way. When the time came, as Rodney Harris has described, to undertake what was called a 'special medical development' (SMD) to explore the application of new genetic probes in the service context, it became obvious that that organization of genetic centres – regional genetic centres as they're called – was not only fitting but necessary to do that kind of work. It was a wonderful confirmation of the ripeness of what had evolved.[136]

---

[133] Sir Donald Acheson (1926–2010) was chief medical officer, Department of Health (1983–91). See Dean (2010) and biographical note on page 113.

[134] Department of Health and Social Security, Scottish Home and Health Department, Welsh Office, Standing Advisory Committee (1972).

[135] For an earlier volume on genetic counselling, see Fuhrmann and Vogel (1969).

[136] Dr Ian Lister Cheese wrote: 'SMD was led by Professors Peter Harper, Rodney Harris and Marcus Pembrey, and included Drs (now Professors) Angus Clarke and Dian Donnai and Drs Helen Hughes, Helen Kingston and Maurice Super.' Note on draft transcript, 19 August 2009; see also note 117.

I think throughout those years in the DHSS, much of our attempt to do rather more in respect of the NHS and its support for clinical genetics was overshadowed by a general unease about abortion.[137] Ministers have always adopted, as you know, 'a neutral stance' over abortion. And it's unfortunate that genetics, indeed the very word, was always closely linked with abortion. But fortunately there was another aspect. A number of people have spoken about the link of genetics and paediatrics and the origins of many geneticists as paediatricians, and it was easy to look at genetics in the context of reproductive health and the health of children, and in doing so to put it into a public health context. You recall that every year the CMO would produce a report on the state of the public health to the Secretary of State. This allowed us to insert, as it were, current preoccupations, current issues and if not actually to give them wide publicity, at least to put them on the record.[138]

Towards the end of the 1980s were a number of profound changes, all of which would have touched you. One, of course, was the establishment of the Human Fertilization and Embryology Authority.[139] That was immensely sensitive. In fact, it was so sensitive, I remember, that the work we were trying to do under

---

[137] Dr Ian Lister Cheese wrote in 1997: 'The development of policy in this field was shadowed by the continuing political and public controversy that abortion arouses....Even the clarification of desired outcomes; for example, to restore reproductive confidence, a phrase coined, I believe, by Marcus Pembrey, has not deflected that sensitivity. It cannot be too often stated that it is not the primary aim of genetics practice to reduce the birth prevalence of inherited disease, although there is good evidence that it is a consequence of informed decisions that individuals and couples make.' Lister Cheese (1997) unpublished: page 3, which will be deposited along with other records from the meeting in archives and manuscripts, Wellcome Library, at GC/253. See also, for example, Polkinghorne (1989).

[138] See, for example, Department of Health (1989, 1992a, 1993).

[139] The Human Fertilization and Embryology Authority (HFEA) was created in 1991 as a statutory body under the Human Fertilization and Embryology Act 1990 to license and monitor UK clinics that offer *in vitro* fertilization and donor insemination treatments, and all UK-based research into human embryos, as well as regulating the storage of eggs, sperm and embryos, including a database of every IVF treatment and one of every egg and sperm donor. See www.hfea.gov.uk/25.html (visited 21 January 2010). Department of Health advice on the disposal of human tissue from pregnancy loss before 24 weeks' gestation was contained in HSG(91)19 (*Health Service Guidelines – Disposal of Fetal Tissue* (12 November 1991)), replaced by the HFEA's *Code of Practice*, and subsequently the Human Tissue Act 2004 (for England, Wales and Northern Ireland only), which set out a new legal framework for the storage and use of tissue from the living and for the removal, storage and use of tissue and organs from the dead. This includes tissue following clinical and diagnostic procedures.

David Weatherall's prompting, to set up a gene therapy advisory committee[140] was objected to in the department because it might get in the way.[141] It was as sensitive as that. However, there were a number of helpful opportunities that came our way almost coincidentally. Again, Rodney has mentioned one: clinical audit. Medical audit first, clinical audit later, became the fashion. Money was made available. Rodney took the opportunity to set up what he called CEGEN, the National Confidential Enquiry into counselling for genetic disorders by non-geneticists.[142] In a sense, it wasn't an audit, it was a way of setting standards in retrospect. It did a number of necessary things to put a clinical service and those who practised it onto the most secure footing possible. Another opportunity arose in 1990. The Health Select Committee looked at maternity services and childbirth.[143] One of the items they considered was preconception care.

Opportunities like that in the department allowed one to declare an interest and to bring in the evolving, or evolved services, in a way that was pertinent to the interests of the committee.[144] And when the Select Committee says something, makes a recommendation or a comment, it can't be ignored. You know that very well from recent experiences with the Lord's Science and Technology

---

[140] The Gene Therapy Advisory Committee (GTAC) was established in 1993, following the recommendations of the 1992 Clothier committee report on the ethics of gene therapy (Department of Health (1992b), Sir Cecil Clothier QC, chair; see also www.galtoninstitute.org.uk/Newsletters/GINL9306/ethics_of_gene_therapy.htm; Taylor and Lloyd (1995)). All gene therapy is considered as research, and recruitment of patients into research trials is covered by strict rules set out by the GTAC. The primary concern is whether the research proposal meets accepted ethical criteria for research on human subjects. The GTAC reviews take account of the scientific merits of the research, and its potential benefits and risks. The safety and welfare of patients is of paramount importance. See http://genome.wellcome.ac.uk/doc_WTD021011.html (visited 21 January 2010).

[141] Professor Sir David Weatherall wrote: 'My guess is that one or two areas in this meeting will come over quite well, but others, like ethics and like issues will probably need an awful lot of work at your end. I suspect that some of these topics could well do for a Witness Seminar on its own. I was thinking of bioethics because there was no time to mention all the fascinating events that led up to the Government recommendations on gene therapy and later on embryo research and so on. Although I hate to say it, but I think this country has done better than almost any other in finding its way through the complexities of modern biology and it might be quite a good topic for you in the future.' E-mail to Professor Tilli Tansey, 29 September 2008. See, for example, Human Genetics Commission (2006).

[142] Harris and Harris (1999); Harris *et al.* (1999).

[143] House of Commons, Health Committee (1991).

[144] Department of Health (1992c).

Committee.[145] So once again we were able to draw the attention of the NHS to the importance of clinical genetics services in the context of preconception care and maternal health and so on.

There was another inhibition, another restriction. At that time, the Department refrained from giving any kind of specific instructions to the NHS. However, the CMO was able to write a Professional Letter on clinical genetics services, and to draw the attention of people to their importance in the developing NHS.[146] I needn't go on beyond that but, as you know, genetics gradually became, more and more, a public issue, and in many ways a threatening issue. Those of us who served the Clothier Committee will remember phrases like 'Frankenstein's monster' being used. It was a most absurd state of affairs, but that was the atmosphere at the time. I think that has passed now.

I'd like to go back to one thing – I'm just recalling it. After the Special Medical Development, which I think ended in about 1986/7, it seemed sensible to produce an interim report before the full evaluation of the development had taken place.[147] It was actually rather a good thing to do because the final report didn't appear for another four or five years.[148] But I remember speaking with Rodney, and to you, Martin, and to Peter Harper and Marcus Pembrey, to see

---

[145] House of Lords, Science and Technology Committee (2009).

[146] Department of Health, CMO and CNO (1993).

[147] Dr Ian Lister Cheese wrote: 'The interim report (Department of Health and Social Security (1987)) was not published formally but was made available to participants. Copies were lodged with the Legal Deposit Libraries. In evidence to the Science and Technology Committee in 1995, the CMO outlined action being taken by the Department of Health in relation to education and training for clinical practice. As a first step the Department drew the attention of NHS education advisers to an educational video "Chance to Choice", supplemented by a booklet, prepared by the Institute of Child Health, London, under the leadership of Professor Marcus Pembrey. A personal view from the Department of Health was contained in an appreciation given by me at a Festschrift to mark the retirement of Professor Rodney Harris in 1997. Many aspects of this personal recognition apply no less to the other leading figures in this field.' Note on draft transcript, 19 August 2009. A copy of the Interim Report, the professional letters and Dr Lister Cheese's appreciation will be deposited along with other records of the meeting in archives and manuscripts, Wellcome Library, London, at GC/253.

[148] Beech et al. (1994); House of Commons, Science and Technology Committee (1995, 1996); see note 120; Brock (1990). Four Scottish university medical genetics centres formed a consortium in 1985 to provide a DNA-based service in prenatal diagnosis, carrier detection and predictive testing for a range of Mendelian disorders. Members were: A W Johnston, N Haites, K Kelly, W J Harris, C Clark (Aberdeen); M W Faed, M Boxer (Dundee); A Curtis, L Strain, L Barron, M Mennie, J A Raeburn (Edinburgh); M Connor, M A Ferguson-Smith, G Lanyon, S Loughlin, E O'Hare (Glasgow). See also Kelly (2002).

what lessons we could draw from it that were persuasive. There were two that I remember now; there were certainly others. The first was, of course, that the new probes allowed much more informative data to be given to patients in counselling. It had one important consequence – that people who would not have risked pregnancy became more confident to do so. I don't think enough was made of that at the time. The other thing that came out of it was the way people sought counselling. I can't remember whether it was part of the SMD or not. It was something like this: 200–300 people at primary risk were written to and offered counselling.[149] A small minority responded in the first instance, and yet over the succeeding weeks or months a majority self-referred or sought advice. The point was that there was a growing awareness of what was on offer. And it brought the prospect of a huge demand for genetic services of the kind you deliver as the possibilities of advice, of information, grew. That presented a real problem. Could genetic services grow in such a way and fast enough to meet that potential demand? I think it was a very important issue because it meant that one had to capture the interest of the patient bodies of that time and say to them: 'Don't ask for too much too soon.' It couldn't be delivered.[150] There are many more aspects, though.

Bobrow: Extremely helpful. I think we were very fortunate to have an understanding ear within the DHSS at the time to translate what was going on into terms that were suitable for internal consumption. I think many of us knew that at the time, but you've confirmed it. Are there other questions or comments on that? I wonder if I could pick up briefly the question of how screening services were developing in parallel with this. And by screening I'd like to include not only the haemoglobinopathies,[151] but also the Down's and neural tube screening.[152] Do you think that those developments would have gone ahead anyway much as they did if there had not been a growing subspecialty, later becoming the specialty, of clinical genetics? Or did the two feed off each other? Nobody knows.

---

[149] See data provided in unpublished interim report (Department of Health and Social Security (1987): Appendices A and B), which will be deposited along with other records of the meeting in archives and manuscripts, Wellcome Library, London, at GC/253.

[150] Dr Ian Lister Cheese wrote: 'The Genetics Interest Group (GIG), an umbrella organization newly formed in 1989, understood the position and pressed its interests and concerns keenly but responsibly.' Note on draft transcript, 19 August 2009.

[151] See note 121.

[152] Spencer *et al.* (1993); Wald *et al.* (1998); Harris *et al.* (1999); Ferguson-Smith (2008).

Harris: I don't know either. I think the thing that is striking is that in the survey of 31 European countries, the UK and a few others like Holland stood out because these sorts of things went ahead, because there were clinical genetics (and scientists working with them) to push them.[153]

Skirton: Could I make a comment from working within an NHS trust at the time when screening for Down syndrome was introduced? The genetic services – the regional genetic service – and genetics *per se*, I think, had very little to do with that, and it was pressure from women who were saying: 'My sister went to Bart's and she had a test, and we want it.' And so it was introduced in a rather unthought-about way – I know this for sure in Somerset, at least. So, I would have said – and actually what happened – is that the genetic services were brought in afterwards to pick up some of the pieces, because it was introduced in a very haphazard way. That's my experience.

Dennis: I was the regional clinical geneticist in Southampton, covering Wessex, in the late 1970s and early 1980s when a lot of screening developments were taking place. I think I felt that I had a lot to say about screening but nobody was asking me, and it took a lot of pushing from the genetics people to get a voice. With the help of Pat Jacobs, we had a regional committee convened eventually to discuss Down's screening. And the obstetricians and other people involved in actually delivering screening didn't seem inclined to look to their local geneticists for help.

Burn: I think this is a good illustration of how things can go different ways in different places. Derek Roberts set up an AFP testing service for spina bifida in one of our district hospitals back in the early 1980s. In fact, we came across the association of hCG independently of Bart's, and so in 1986 we started doing double testing for Down's ourselves.[154] I decided to take it over, because then I could incorporate the cytogenetic costs into the price of the biochemistry. So we introduced a regionalized service, but, in fact, it was for only half of the health districts. We negotiated district by district. It was £14.20 a test, I remember it vividly. We actually controlled Down's screening all the way through until it became a national screening test run by our regional screening lab around 2004.

---

[153] See note 122.

[154] Professor Sir John Burn wrote: 'The discovery came out of our NHS scientist Irene White working under our senior lecturer in human genetics with whom I worked. The former had little interest in academic publication and the latter failed to understand the urgency, so the work was not submitted to a journal, but we deployed it quickly into our clinical practice. By the time we had more data on its relevance, the "Barts test" had been released.' E-mail to Mrs Lois Reynolds, 22 March 2010; see Wald *et al.* (2003).

Berry: Guy's was rather the other way round from Southampton, in that we very much controlled the screenings, certainly for Down's, because we could only handle a certain number of samples each week and there were strict criteria that obstetricians had to fit in with. The actual AFP assay was all done at Guy's. Certainly we would see the obstetricians, who were always relieved to hand over any difficult counselling or difficult results or anything. So it was very much a partnership, and I think probably the Guy's group drove the service to some extent.

Emery: I think in Scotland in the late 1960s–early 1970s we had three screening programmes going on. They were screening for phenylketonuria in Glasgow in newborns (1967) (this was nothing to do with genetics; this was, I think, by the Public Health Department).[155] We took it up later using the same technique when screening newborns for Duchenne muscular dystrophy.[156] But it was David Brock who started AFP screening in Edinburgh, and then it spread all over.[157] So two tasks were stimulated by geneticists, CPK and AFP, and one stimulated by the Health Board with PKU.[158]

Bobrow: I guess, for completeness, from my memory, Nick Wald was in Oxford and did much of the basic work.[159] This came out of a collaboration between genetics and Nick Wald and was driven clinically from the Oxford genetics department. It wasn't all done there, but it was controlled there. So, quite a mixed pattern across the country. Now, dysmorphology.

Donnai: We didn't invent dysmorphology (the study of malformation). You have to remember that within pathology and anatomy, there had been many scholarly descriptions of syndromes in the old literature that we now recognize. In the UK, when I first began in genetics, people like Cedric Carter were undertaking epidemiological studies about individual malformations like cleft palate and Hirschsprung disease, etc., and deriving risk figures which are very good and still

---

[155] Consultant Paediatricians and Medical Officers of Health, South-East Scotland Hospital Region (1968).

[156] Duchenne muscular dystrophy (DMD) affects 1 in 3500 newborn males, causing progressive weakness of the muscles. The test involves the measurement of an enzyme (CPK) using the neonatal blood spot collected for the phenylketonuria test, part of the routine newborn bloodspot screening. A newborn screening programme for DMD has been running in Wales since 1990, evaluated (1990–98) and health authority-funded from 1998. Bradley *et al.* (1993).

[157] Brock (1990); see note 152.

[158] National Medical Consultative Committee, Working Group (1986).

[159] See note 154.

The Hospital for Sick Children

Great Ormond Street London WC1N 3JH

PATRON: HER MAJESTY THE QUEEN

TELEPHONE: 01-405 9200

TELEGRAMS
GREAT LONDON WC1

CHAIRMAN:
MRS. A. CALLAGHAN

HOUSE GOVERNOR:
R.G.B. MILCHEM, F.C.I.S., A.M.B.I.M.

JB/SL

13th October, 1981

Dr. Dian Donnai,
St. Mary's Hospital,
Dept. of Medical Genetics,
Hathersage Road,
Manchester M13 0JH

Dear Dr. Donnai,

The next dysmorphology meeting will be on Friday January 15th, beginning as usual, with lunch at 12.30 pm. As we discussed, participants will be asked to contribute £1 each to help cover catering and stationery costs. Room 203, second floor, in the Institute, has been reserved for the meeting.

The special subject will be the Russell Silver Syndrome. Mike Preece, from the Growth and Development Unit, will review the series of cases seen in their department.

The case discussion will follow Dr. Preece's presentation, beginning at about 2.00 pm. and continuing until 4.30 pm. As before, you are invited to bring slides of known and/or unknown syndromes. Each speaker has 10 - 15 minutes. Please complete a brief typed resume, with one reference, on the attached sheet and bring this to the meeting.

I enclose a copy of the case reports from our September gathering.

Yours sincerely

John Burn MB MRCP
Hon. Senior Registrar In Clinical Genetics

. . . .

Figure 15: Dysmorphology Club meeting invitation, 1982.

in use today. But in terms of the individual syndromes, I don't think that was Cedric's focus. The term dysmorphology was probably coined by David Smith in Seattle, who was working in the 1970s.[160] In the US there were a series of conferences supported by the Teratology Society and the March of Dimes.[161] David Smith was an excellent teacher and had quite a lot of fellows. He died tragically early in 1981 and his ex-fellows set up a series of conferences that are still going on, the David Smith workshops on malformations and morphogenesis.[162] I was the first person from the UK to attend this meeting in 1984.

But things were going on in the UK as well and indeed in 1984 I began the biennial series of Manchester birth defects conferences, called the Manchester Dysmorphology Conferences from 2008, which are still hugely oversubscribed and very international now. This is something that has been one focus of dysmorphology in the UK. But dysmorphology in the UK was built on the interest and enthusiasm of a number of individuals, and foremost among these were Michael Baraitser and Robin Winter, initially of the Kennedy-Galton Centre and then at Great Ormond Street. People forget that Robin worked for a long time at the Kennedy-Galton. I remember Michael, Robin and I standing in corners at CGS meetings with slides looking at pictures of patients in the late 1970s. The Dysmorphology Club, which is still held three times a year at Great Ormond Street, began as a lunchtime meeting when Cedric was still there, in the library, I think in 1980. I travelled all the way down from Manchester, which was quite a long journey in those days, three hours plus, for this one-hour lunchtime meeting and then went all the way back. I eventually persuaded them that it should be at least half-a-day. So in 1981, I think, when John Burn was there, we had this half-day meeting and wider participation from other centres was encouraged. By 1982 it was so popular that this was a whole-day meeting and I've got records of John's invitation to the meeting asking for £1 from all participants for tea, coffee, biscuits and photocopying (Figure 15).

That tells you how long ago that was. I think the Dysmorphology Club, as well as being useful for the diagnosis of individual patients, was extremely important in defining new syndromes by people being extremely collaborative, setting up research collaborations, and in fact some of the major studies that have been

---

[160] Smith (1966, 1977).

[161] See www.teratology.org (visited 6 May 2009) and for details of past annual meetings of the Teratology Society, see www.teratology.org/archivedMeetings.asp (visited 6 May 2009). See also www.marchofdimes.com (visited 6 May 2009).

[162] See http://dwsmith.org/ (visited 6 May 2009).

Clinical Dysmorphology

VOLUME 1 NUMBER 1 JANUARY 1992 ISSN 0962-8827

CONTENTS

EDITORIAL — Page 1

PAPERS

AINSWORTH, J.R., SHABBIR, G., SPENCER, A. and COCKBURN, F. — 3
Multisystem disorder of Punjabi children exhibiting spontaneous dermal and submucosal granulation tissue formation: LOGIC syndrome

TEMPLE, I.K. — 17
Char syndrome (unusual mouth, patent ductus arteriosus, phalangeal anomalies)

HOULSTON, R. and MACDERMOT, K. — 23
Acrorenal syndrome: further observations

PRIDMORE, C., BARAITSER, M. and LEONARD, J. — 29
Ramon syndrome with diabetes mellitus and vascular skin lesions in two sibs

FROSTER-ISKENIUS, U. and MEINECKE, P. — 37
Encephalocele, radial defects, cardiac, gastrointestinal, anal, and renal anomalies: a new multiple congenital anomaly (MCA) syndrome?

MEINECKE, P., BÖNNEMANN, C.G. and LAAS, R. — 43
Microgastria-hypoplastic upper limb association: a severe expression including micropthalmia, single nostril and arhinencephaly

YOUNG, I.D., ZUCCOLLO, J.M., BARROW, M. and FOWLIE, A. — 47
Holoprosencephaly, telecanthus and ectrodactyly: a second case

NORMAN, A.M., READ, A.P., CLARK, A., HASLAM, J. and DONNAI, D. — 53
Fetus with unbalanced translocation involving chromosomes 2 and 11

CONFERENCE REPORT — 57
First Robert J. Gorlin conference on human dysmorphology. Minneapolis, October 14th–15th 1991

Clinical Dysmorphology is published by
Chapman & Hall, 2–6 Boundary Row, London SE1 8HN, UK

Figure 16: Contents of the first issue of *Clinical Dysmorphology*, 1992.

done on individual syndromes were started from that group. I think it's the perfect example of a cooperative. It only cost £1; it wasn't official in any way, and it still isn't official in any way, yet it's enormously popular.

Robin and Michael, meanwhile, were working on another major contribution to dysmorphology, that is the dysmorphology database, and the prototype of that was distributed informally from about the mid-1980s and then published by Oxford University Press in 1990. It's a tool for experts and it's in use worldwide still. Following Robin's death in 2004, Michael has continued to ensure that this is up to date and that excellent standards are maintained. He renamed it the Winter-Baraitser Dysmorphology Database and this is now distributed by Michael.[163] There's also the journal, *Clinical Dysmorphology*, which is in the colour that Robin chose and we always used to call 'dysmorphology pink' (Figure 16). After Robin died, Jill Clayton-Smith joined me, so Jill and I are the editors now, although Michael still supports us.

I think there have been a number of centres in the UK that have trained people in dysmorphology and have populated many other centres as well: London, Great Ormond Street and Guy's being the main ones; and then there's Manchester; and then in Cardiff, Helen Hughes was recruited by Peter Harper in about 1986. Obviously John Burn went from Great Ormond Street to Newcastle and Judith Goodship joined him there. Those have been the main centres where there has been a lot of dysmorphology research as well as training.

The last point I'd like to make focuses on the patient cohorts that have been built up over the years that are national patient cohorts, and have been often in collaboration with patient groups. There has been incredible progress in identifying the genetic mechanisms underlying many of these syndromes that we've always recognized as families of syndromes linked by phenotypic similarity. It's very gratifying that by understanding the pathways in which the genes that are mutated operate that our clinical 'eyes' are now being given some credence. We're even, for some of these disorders, beginning to look at the possibility of modifying the disease process and even treatment. So, I think dysmorphology has moved on from when I first went into genetics. I remember somebody saying to me: 'What does dropping out of real medicine feel like?' I feel that we've all been very privileged to be able to be involved in the delineation of disease, the study of natural history, and then understanding of the mechanisms and beginning to look forward to the treatment of some of the most intractable and rarest diseases.

---

[163] See www.lmdatabases.com/about_lmd.html (visited 6 May 2009).

Burn: Apart from saying that Dian Donnai has been absolutely influential throughout the last 20 years in keeping dysmorphology focused as a clinical aspect of our specialty, I think it is worth noting historically that the beginning of the 1980s was a very important time for the specialty because that was when we grabbed hold of paediatrics again. I joined having completed a paediatric training. People like Peter Harper had established our presence in general medicine, but we hadn't secured our base as a paediatric subspecialty. It gave us an area that we as a group of clinical consultants could genuinely own. We could go to any city, any hospital within the country and we walked tall because we could recognize Rubinstein Taybi syndrome and they couldn't. That still is an asset to clinical geneticists and it's one of the reasons why we won't die out, I suspect, because you need someone who focuses on doing rare problems enough so that they can get their eye in and keep it in, and you can't actually get rid of that. I should say that Michael Baraitser was a source of encyclopaedic knowledge and was the most wonderful man to work with, but he was an adult neurologist. I remember having to teach him how to examine children; he thought they always cried when you examined them. I showed him how to do it without actually making the child cry and it was a great illumination to him.

Pembrey: I think we've illustrated how clinical dysmorphology took off from that first meeting in 1980, but there were some real politics behind the scenes, which largely fell to Michael Baraitser and me. When we said we were going to do that meeting, every single paediatrician in Great Ormond Street felt they should be invited. We only had a small space, and they all came and wanted to know why we couldn't change the time because it clashed with their clinics and so on. So, the paediatricians realized what was happening and we had to decide how we were going to deal with it. We didn't want it to be elitist, by invitation only, so we decided we would talk dysmorphology in fine detail with no concessions for people who didn't know what we were talking about. That continued, and after three meetings, all the general paediatricians stopped coming. It defined the subspecialty, and it worked a treat.

Bobrow: Yes. This is the secret of much good government. Excellent. Let's go from that to the subject of how the counselling side of genetic counselling developed.

Skirton: We have less history, I think, but nevertheless it's very much entwined with the history of medical genetics and our medical colleagues. As we've heard, in the early days – even in the 1960s but certainly in the 1970s – there were a few health visitors or social workers who were working as part of a clinical genetics team or alongside medical geneticists. But we date our profession from 1980 when the association was formed with nine nurses and one social worker: it was named the Genetic Nurses and Social Workers Association (GNSWA) and that name stuck until we joined the BSHG and took the opportunity to change our name, partly because we'd sat in meetings with John Burn for many years while he laughed at the name. So we had wanted to change the name for some time. During the 1980s the association gradually grew from the original ten members to roughly 100 members in about 1990. It's true to say that there were some initiatives that helped the growth of the genetic nurses and later genetic counsellors. One of those was the course established by Mary Rogers in Cardiff under the Welsh Nursing Board, which was a module in genetics nursing that was the key to many people developing the skills they needed to work more autonomously in the field. Up to that time, we had all worked and learned on the job, by a process of osmosis, working alongside our very skilled medical genetics colleagues. But that course defined a new type of skill for genetic nurses, alongside those we already had. As I mentioned, we had a lot of support from the Clinical Genetics Society and we started a scientific meeting in the late 1980s. At the time there was no such thing as a genetic counsellor and there is someone sitting to my right – Lauren Kerzin-Storrar – who is virtually responsible for single-handedly introducing that breed of person to the UK. Lauren trained in the US, came to Manchester and was very innovative. She introduced a Masters' degree in genetic counselling that has since trained many counsellors.

At that time there were changes in nursing which meant that nursing was becoming more professionalized, and nursing practitioners were becoming more autonomous. There were also these counsellors who had a Masters in genetic counselling but no other professional background, and we started to think about regulating this new profession. Nurses had their professional statutory guidance, but the genetic counsellors who had done the Masters did not. We felt that there were concerns about setting standards for patient care. During the 1990s, Penny Guilbert, who's not here today but has had a huge part in this, started an education group, which I chaired for a long time. And we started to look at education, the role and the scope of working of genetic nurses and counsellors. We four here – Chris Barnes, Ann Kershaw, Lauren and myself – were in the education working group and first we did a study where all the genetics nurses and counsellors and the medical geneticists in the UK were asked what they thought the role and scope of practice for genetic nurses was.[164] That was quite powerful, because not only did we have the views of nurses, we had those of our medical colleagues. We did this study in 1996 and I have to say, it was wonderful to get the responses from our colleagues who demonstrated their faith in the way that we were working and developing. It was the impetus for us to go on and develop more autonomous roles. At that time, most of us were seeing some patients on our own, as well as doing pre- and post-clinical work with our colleagues. There were occasional pockets where nurses were still getting the doctors' coffee and changing the sheets on the examination table – and that is true – but this study helped us to change those outliers. We did publish a couple of papers on our role and the educational standards, and we started to think about registration.[165] In 2001 the Association of Genetic Nurses and Counsellors voted to have a registration system, a voluntary system, whereby people demonstrated their competence through a portfolio. That system has been highly successful to the point where now we are actually one of the professions applying for registration through the Health Professions Council, and we hope that next year (2009) that will become a reality for us, and it won't be a voluntary system any more, but one that truly establishes our profession among the big boys.[166]

Bobrow: Lauren, do you want to say anything?

Kerzin-Storrar: Not so much in the historical sense or about the evolution of the profession, but something about the kind of ethos and scope of non-medical colleagues in the clinical genetics team. We've talked about some of the very early pioneers like Kathleen Evans at Great Ormond Street and Audrey Tyler in Cardiff, who were from social work backgrounds. But I think it's fair to say that in the early days those individuals and other health visitors were often employed on research money with the aim of gathering extended pedigree information and to obtain research samples. It started out in that way. The role then moved to encompass provision of psychological support around pregnancy issues and diagnoses, and now it has evolved to a combined role which includes the previous roles, but also autonomous genetic counselling for individuals where the diagnosis is established. It now includes discussion of the genetics and decision-making around testing with patients and their families. That is important to say.

---

[164] Skirton *et al.* (1997, 1998).

[165] Fine *et al.* (1996).

[166] See Appendix 2, pages 87–8.

The issue about the term 'genetic counsellors' has been something interesting, because I've worked now for nearly 30 years in the UK in this field, and for a long time – I'm looking at Peter Harper because we have had a lot of discussions about this over the years – for a non-medical geneticist to call themselves a genetic counsellor was a difficult issue. And my sense was that when you as medical geneticists were still establishing yourselves as a recognized subspecialty, you rightly thought you were doing genetic counselling – and you do do genetic counselling – and that's what you, in many ways, saw yourselves as, in addition to being diagnosticians. But as you've become more established, I think people became more comfortable to call a different group of individuals 'genetic counsellors'. In past editions of your *Practical Genetic Counselling,* Peter, there was always a little paragraph – and I used to go and look at it with each edition to see where you were up to in your thinking – and it's evolved quite a lot.[167]

Skirton: May I add to that? Virtually every genetic nurse or counsellor I have ever known has confessed that, for the first two or three years of their practice, they didn't let Peter Harper's book out of their sight for more than about five minutes because we relied on it so much.

May I share one anecdote that reflects the naivety of the Genetic Nurses and Social Workers Association in the early days? We didn't know what we were doing and it was decided that a constitution should be established for this new association. But no-one had ever done that before and one of the members of the group was a member of a sailing club and she went and got the constitution for the sailing club and they crossed out 'sailing' and put 'genetic nursing'. We lived with that constitution until the BSHG was formed.

Mrs Ann Kershaw: There are two people here that have certainly had a huge effect on my career in genetics, and the reason I'm in genetics is because, as a health visitor, one day I went to a talk and there was somebody called Clare Davison there who was so enthusiastic about this thing I had never heard of called 'clinical genetics', that I thought, 'I have got to do this.' So after many years working in the department of genetics in Cambridge and enjoying my role, but feeling that we perhaps hadn't quite developed, Martin Bobrow came along and I remember him sitting down and talking to me, and we sat in a clinic and he said: 'I think you're all far too well qualified to be sitting in a corner.

---

[167] See the section on 'The genetic counselling clinic' in Harper (1981): 114–15 and Harper (2004): 138–9.

Go out and see patients.' And that's what we started to do. Counsellors in our department now see 40 per cent of referrals and work very happily alongside our medical colleagues; the thing that has grown is the mutual respect that we have for each other, and that's something that I value and I think benefits our patients.

Ms Chris Barnes: I'm the fourth member of our merry band and I want to acknowledge the Department of Health. In the 2003 Genetics White Paper,[168] it was acknowledged officially that a training scheme for genetic counsellors was needed, something that the four of us, together with our other colleagues, had striven very hard to establish. The funding for that scheme enabled us to appoint 43 trainees, who have either graduated from that scheme or are still in training, and this has been an enormous help. We have had many productive conversations with the Department of Health about the importance of training genetic counsellors and the future of genetic counselling. Without the Department of Health's help we wouldn't have the strong position that we do now.

Professor Angus Clarke: I think it is interesting to think about what has happened in the UK and compare that with what's happened in European countries, where the role of genetic counsellor is hugely varied; in some countries it is very well developed, but in others it scarcely exists. In some countries the one or two genetic counsellors there have to develop the role against a rather hostile bunch of medical practitioners. It would be interesting to think through the factors that have led to those differences.

Harper: Thanks to Heather Skirton and others for being so charitable. I look back at the early editions of my own *Genetic Counselling* book sometimes and think: 'Good heavens, did I write that?'[169] I have to say my attitudes have changed a lot. I am particularly interested and heartened by what Heather has said, because I myself have always felt that the links between genetic counsellors and nurses, and the clinical geneticists have always been strong and mutually supportive. But, then I wondered: 'Well, is it just me thinking that?' I started reading quite recently about the experience in the US, where there was a very bitter division in the late 1970s. I hadn't realized how bitter until I read about it; a very exclusionary – mutually exclusionary – attitude was taken up by both

---

[168] Department of Health (2003). The White Paper covered arrangements in England, with those in Wales, Scotland and Northern Ireland outlined in an appendix.

[169] See page 71.

camps. I hadn't realized this had even happened and I thought: 'My god, I'm sure we didn't have that same problem in the UK'. So, I am relieved to hear Heather say that we didn't.

May I put in a plug for records? I think the early records of groups like the Association of Genetic Nurses and Social Workers and these other groups are important, particularly those from the stages before they got big or too formalized. The Dysmorphology Club, again I hope people will make an effort to preserve these records.

Hodgson: I'd like to bring in cancer genetics here, because it doesn't get a heading, as I've noticed, but it is something which has become a very large part of clinical genetics in the last 20 years or so. Certainly, in the 1990s cancer genetics was hardly known, and then people like Bruce Ponder[170] and Joan Slack[171] started clinics and then it mushroomed, as everybody knows. Of course, I was interested in what Angus Clarke said because I did a survey of cancer genetics services in different European countries and the services were very much enhanced in the countries where nurses were able to practise and help with the service, because it's the sort of service where you need a lot of people and certainly nurses and other non-medical people can play an enormously important part in developing this kind of service.[172] So I would say that that's been very important in this major development in clinical genetics.

Skirton: That's very interesting, Shirley, and I was going to mention that, in the survey we did in 1996, we asked medical colleagues and nurses where they thought development of genetic nursing would lie, and cancer, which was just looming on the horizon as a very big task, was cited as an area where genetic nurses could play a major role.

I want to go back to what Peter Harper said because I think when we oldies sit around and talk about how our careers have developed, we always go back to the central component of the development of a trusting relationship between individuals, between a doctor and a nurse working in a department that enabled that person to develop and go on to become a leader in the profession. I think perhaps that is what differs in Europe and some other countries, but it is certainly true for us in the UK.

---

[170] Ponder (1987, 1991).

[171] See, for example, Murday and Slack (1989); Iselius *et al.* (1991); Houlston *et al.* (1991, 1992).

[172] Hodgson *et al.* (1999); Hodgson (1999).

Bobrow: Peter, you were going to explain why cancer genetics is not on the agenda.

Harper: We didn't forget about cancer genetics, but felt that not only was it pretty recent but that it was a very big topic and that to cover it would require a completely different community and set of people. So we thought we'd leave that for Shirley Hodgson or somebody to suggest as a separate Witness Seminar.

Donnai: Angus Clarke raised the issue as to the reason why genetic counsellors and clinical geneticists and most of us work in relative harmony most of the time. Firstly, I think that is because we have a socialized healthcare system, and secondly we have regional genetic services. One of the few other countries where this exists is the Netherlands and I think services have similarly thrived. Whereas in the US, of course, there are pockets of great brilliance in genetics and genetic research, but many of my colleagues in genetics in the US are one-man bands. People there call themselves dysmorphologists and have no connection whatever with genetics. I never call myself a dysmorphologist, I call myself a clinical geneticist, because that is the family to which we belong and I subspecialize in that particular area. Even in France, although there are attempts to try to look at things country-wide – they've got excellent services for people with rare diseases and investment in research into rare genetic disorders – they still haven't got the same approach to regional genetic services. In the UK, there is cooperation between centres, and collaboration in research studies, a willingness to let other people have your patients and samples from your patients or to be part of a multicentre study – these are part of the reason that UK genetics has been so successful.

Bobrow: Interesting comment with which I will not argue just yet. What did I mean? That's not fair, is it? I think there are lots of examples within the NHS of other parts of the medical profession who still today regard anyone who doesn't have a medical degree as pretty subservient. I think we got over that as a community a couple of decades ago and I think that we were ahead of the game.

I am going to go back to the agenda. We are going to skip the role of lay societies, with some regret, because as it turns out, although several people were invited and slightly fewer agreed they would be here and on the day – for one reason or another – there aren't any attending. So we will note, and I don't think there's any argument about it, that there's been a lot of interplay between the professions and lay societies in the field of inherited diseases. They've been extremely active and incredibly helpful; it's been a very productive relationship and we all know who should have been here to talk about it, and you'll try to somehow or other to work it in when you write, about the ethical and social dimension in clinical genetics.[173]

Clarke: It is usually the ethical and social bits that drop off the end of the agenda. Obviously, in a sense, it would need a much bigger group of people to try to represent what has been discussed regarding the social and ethical issues surrounding genetics, because it has involved so many more people than those within the profession. I will try to mention three or four themes or issues that have cropped up and come out with a few statements relating to each and people can carry on discussing it over a glass of wine afterwards. The areas I'm going to think about are: reproduction, very generally; testing individuals with an individual focus, that's another area; family issues, that's a third area; and then the transition between research and service, as a fourth. There are a lot of other ways one could do it.

There are a lot of issues around reproduction – I feel I came into this halfway through – but as I remember, if you look at some of the literature in the 1970s and 1980s, there were a lot of papers about reproductive population screening which focused very much on population outcomes, cost of screening versus cost of care and things like that, which I think caused quite a bit of offence outside the medical community and made quite a few people within it uneasy.[174] There was an effort to shift terminology away from cost rationales for screening through into individual and informed choice, and to some extent that has been a cosmetic operation. I think that it's not only cosmetic, but there has been quite a significant shift in how services are delivered, if you look back 20 or 30 years. But there was quite a lot of debate, I think, particularly in this service, and the disability rights groups were quite vocal in discussing these topics in a very broad public agenda. I think professionally that led into a desire to emphasize non-directiveness as a way of distancing ourselves from the more public-health driven screening programmes. One can see how that shift towards non-directiveness was a way of preserving one's dignity as a profession and was, I think, synergistic along with the development of the genetic counselling community. I think,

---

[173] See, for example, note 141.

[174] Professor Angus Clarke wrote: 'For the use of "cost of screening is less than lifetime cost of care" as justification of antenatal screening, see Hagard and Carter (1976), which raised the important questions and called for a debate that then failed to happen. See also Boston (1981); Wald *et al.* (1992); Anon. (1992); Shackley *et al.* (1993); Piggott *et al.* (1994); Clarke (1997a).' E-mails to Mrs Lois Reynolds, 11 and 15 March 2010.

to some extent, those things were happening at the same time. Following on from that battle, because it is more or less won, I think the individual focus on reproduction has carried on in other ways: there have been discussions about terminations of pregnancy for less serious conditions, preimplantation genetic diagnosis and for what conditions termination is acceptable. Those things have come along more recently and they've had a much greater focus on what the individual family or woman wants to do, and negotiating that outcome with the professionals rather than stepping back to the population perspective.

Moving on to testing individuals and how that has developed, or been thought about, while I was just beginning to get into genetics – from the beginning as far as I was concerned – there was a very cautious approach to predictive testing, particularly as the early testing for Huntington's disease served as a paradigm for this.[175] This very cautious approach of 'Let's go very gently, see what happens' and to develop the service in that way again fitted in with the genetic counselling perspective. I can see one or two factors that led into that: one was the involvement of the patient support groups and family federations in the mid-1980s. I don't know how big a factor this is, but I think it might be relevant that the clinicians involved in predictive testing for Huntington's disease often came out of a research base where they had tracked families, got on their bikes and gone around and met a lot of family members, collecting blood samples to do linkage studies that were an essential preliminary before this could happen. So, there were a lot of clinicians who had met and got to know a lot of affected, at-risk, and unaffected and married-in spouse members of these extended families. That helped, and I think, perhaps, the relationship with the family support groups reinforced this very calm decision to have a very cautious, measured approach to things. Maybe I'm reading too much into the personal experiences of the clinicians, I don't know. This strand of being cautious has developed since then as well; an obvious example that I've been involved with is genetic testing in children. That was at the Clinical Genetics Society working group reporting the early 1990s, which Lauren Kerzin-Storrar contributed to and that is a topic that still bubbles away on the back burner and hasn't quite gone away.[176] But that's another area where this has carried on, and I suppose

---

[175] See, for example, Craufurd and Harris (1989); World Federation of Neurology (1989, 1993); Tyler *et al.* (1992); Harper and Clarke (1990); Harper (1997); see also Harper *et al.* (1981).

[176] Professor Angus Clarke wrote: 'There were protests by disability rights activists, such as one in London aimed at prominent medical researchers enthusiastic about antenatal screening, carried on by the group of adults affected by Down syndrome, People First.' E-mail to Mrs Lois Reynolds, 11 March 2010. See also Hubbard (1997); Parens and Asch (2000).

you can see it in our collective response – not quite a collective response, but the collection of our individual levels of distress – in relation to some of the inappropriate offers of susceptibility screening and the commercialized medical testing that one sees now.[177] I'm not sure that as a profession we've done very much about that, we've done little bits. But this measured approach to trying to help think through with families when a test is helpful and when it isn't stems back to the mid-1980s.

On to the family dimension, I'm thinking about privacy and confidentiality issues. There were quite a lot of fears and statements in Nuffield Council and RCP committees about how to deal with individuals in families who do not pass on genetic information to their relatives when we think they should.[178] This prompted quite a lot of ethical and legal soul-searching, and there's been quite a lot of discussion about that, some fairly academic, but I don't think it involved the profession and beyond half as much as it should have. But there's been quite a lot of academic work on that and it's been a theme that has cropped up in the Wellcome Trust biomedical ethics research programmes. There have been several projects they have funded on that.[179] One of the responses to that set of issues and concerns has been empirical studies of 'how do family members actually handle communication and pass information on to one another?', which has been a very productive response to tackle the difficult issues, rather than the profession, in deciding to introduce a set policy on forced disclosure and drawing up detailed statements about when to drive round to tell someone that a family member hadn't passed information to them. A different route has emerged there.

Regarding research into service transitions, there are quite a few that one could touch on and several have been mentioned already: cancer genetics is one; the role of the genetic counsellor is another, so many of the early proto-genetic counsellors being research funded.[180] Another is the legacy research

---

[177] Professor Angus Clarke wrote: 'Numerous offers of commercial testing for genetic susceptibility to the complex disorders are available through the internet or face-to-face. These are of dubious worth and the evidence of any clinical utility is minimal or absent.' Note on draft transcript, 11 March 2010. See also Rafi *et al.* (2009); www.hgc.gov.uk/client/Content.asp?ContentId=752 (visited 16 April 2010).

[178] Royal College of Physicians, Committees on Ethical Issues in Medicine and Clinical Genetics (1991b); Nuffield Council on Bioethics (1993); Clarke (1997c); British Medical Association (1998).

[179] Featherstone *et al.* (2006); see also, for example, Forrest *et al.* (2003). See 2004 feature on family communication by Karen Forrest at http://genome.wellcome.ac.uk/doc_WTD022305.html (visited 22 January 2010).

[180] For genetic counsellors, see pages 35, 68–72.

results generated in a project that then ceases, and you go back and look at the workbooks of people and wonder what to do with those results when a family or family member comes along. Also, the drive to proper laboratory accreditation has been quite an important transition. I think the cohorts of families who have been studied for research, particularly the big linkage families (the muscular dystrophies, Huntington's disease and some of the cancers, in the linkage phase of the work), there were often management clinics or at least some sort of management bonus that the families involved got from participation in the research. As that phase of research has gone away, a lot of those clinical structures have lingered on, which I think is quite appropriate.[181] Then we come up against NHS definitions about what counts as a genetic service, to what extent provision of practical support is appropriate for a genetic service, and it shades into coordinating surveillance for those at risk of cancers. A lot of areas, particularly multi-organ cancers, are accepting the coordination of surveillance for people at risk of those conditions. Probably fewer want to take on coordination and surveillance for those at risk of maybe one tissue being involved, and leaving that to the surgeons or whoever. When it comes to the non-cancers, I think we're in disarray and areas are very different and I think we should be working something out. I'll be very interested in people's comments, and I hope I've provoked a few people.

Bobrow: Is anyone provoked? Shirley is provoked.

Hodgson: Yes, I have a quick comment about the role in terms of surveillance for people with cancer-predisposing conditions. It's interesting, because I did a survey of the management of people with hereditary non-polyposis colon cancer in the London regions. We wrote to a lot of GPs and gastroenterologists at genetics centres and asked who they thought was responsible for the maintenance of the follow-up surveillance for these people, and everybody said it was somebody else. Basically nobody thought it was them. I thought that was quite interesting. There were one or two genetic centres that were beginning to think their registers could be used to help with this, but there were an awful lot that didn't feel that the actual responsibility for sending out the appointments for surveillance was their responsibility. It was quite informative.[182]

Pembrey: I think it would be right to mention Gordon Dunstan in this because

---

[181] Professor Angus Clarke wrote: 'These clinics arise and continue or cease for multiple, complex, interacting reasons.' E-mail to Mrs Lois Reynolds, 18 March 2010.

[182] Geary *et al.* (2008).

from his base at King's College London, as the holder of the first F D Maurice chair of moral and social theology, he was engaged fully with the Association of Paediatrics and the Clinical Genetics Society in various ways and was immensely helpful in the discussions.[183] I certainly felt so.

The other thing that I want to flag up is that we've skipped over the Special Medical Development that the Department of Health funded, or we talked about it but perhaps have not drawn out as much as we might have from it. That funding went over three or four years and was then extended and grouped together in a very practical way, while we were working out how we were going to apply these DNA probes, what we should tell, what about the error rates, and so on. Certainly for me, I thought it was the groups getting together and thrashing out things like the objectives of what we were doing. I remember travelling back from Cardiff with Bernadette Modell and trying to find a phrase that countered the pro-lifers' 'search and destroy' tag that they were giving the use of prenatal diagnoses using probes and so on.[184] Modell's beautiful work on the succession of what happened in the Cypriot families with β-thalassaemia, if they didn't know anything, they just went on having lots of children, some of whom would be affected.[185] When they had the counselling but not the prenatal diagnosis, they stopped having babies. But with the offer of prenatal diagnosis the number of pregnancies was restored to the level when they were ignorant of the genetic risk. This idea of restoring reproductive confidence, thinking that through – it's all very obvious now – it was a way of presenting an objective for what we were doing then, among others, which I think was very helpful in moving things forward.

Bobrow: I don't see anyone else rising strongly. It suggests that what you're saying sounds less revolutionary than it might have done a year or two ago. It was an extremely fair summary, I thought. If I move on from the ethical issues, we've heard nothing yet about the interface with primary care. We happen to have one former GP, self-confessed – although I hadn't known about Ian Lister Cheese – and two others, Hilary Harris and Michael Modell.

Professor Michael Modell: Very briefly: I have tried to get primary care engaged in genetics since about 1990. My aim has been to encourage health professionals

---

[183] See, for example, Shotter (2004); Reynolds and Tansey (eds.) (2007): especially 17, 45–7, 75–6, 107–8, 175–6.

[184] See, for example, Christie and Tansey (eds) (2003): 70.

[185] Mouzouras *et al.* (1980).

to view genetics as part of mainstream clinical and reproductive medicine. Until recently, I suspect most family doctors considered clinical genetics to be a very specialized subject, only affecting a few of their patients and therefore of limited interest. GPs have little contact with clinical geneticists. They usually refer one or two patients with a possible inherited condition (or anxiety about their family history) every couple of years. If a genetic diagnosis is confirmed, the patient and family are then likely to be referred on to another hospital specialty (e.g. lipid clinic, paediatrics or haematology, where patients with the commonest single gene disorders are seen). From then on the interaction will be between the primary physician and secondary care specialist, rather than between the primary physician and clinical geneticist.

The narrow primary care view of genetics is slowly changing, partly because of the introduction of the national reproductive screening programmes, the increasing emphasis on family history, the establishment of the NHS National Genetics Education and Development Centre[186] and the work of the small number of GPs with a special interest in genetics. We are beginning to make some headway encouraging GPs, health visitors and community nurses to consider genetics as part of mainstream medicine. For example, the postgraduate curriculum of the RCGP includes a substantial section on genetics in primary care. In spite of that, there still often appears to be a gulf between primary care and clinical genetics. Perhaps the problem lies within primary care, though it is also possible that many (but certainly not all) clinical geneticists have difficulty relating to primary care because there is so much of it and it seems rather remote. Primary care has not figured much in today's seminar.

Dr Hilary Harris: The difficulties don't lie in primary care. Of course, they do a little, but I think there is enormous synergy between genetics and primary care, because we look after families, we look after them for a long time, all the elements are in place, but, as Michael said, the anxieties about the work of genetics are sometimes overwhelming. I think where we have made huge progress – and where I think the building blocks are there – is in the take-up of IT within primary care. I'm quite surprised that that's not come up at all this afternoon, because it's such a major part of actually looking after people. When

---

[186] The NHS National Genetics Education and Development Centre, funded by the Department of Health and located at the Birmingham Women's NHS Foundation Trust, was one of the major initiatives of the 2003 genetics White Paper (DoH (2003)) to ensure that the potential benefits of genetics are realized by the NHS and is intended to improve the understanding of genetics among all health professionals and its role in modern healthcare. For further details, see www.geneticseducation.nhs.uk/about-us/key-aims.aspx (visited 22 January 2010).

you think that in a practice of 5000, where I practised until quite recently, we had every major diagnosis on the computer. I'm quite sure that's so for every practice in the country. It's quite easy to have a register; it's quite easy to look for people, to screen them, to put in prompts, to recall them, all the things which I think are also elements of genetics.

Sadly, thinking back to the 1990s, and again we haven't mentioned the Human Genetics Commission (before that the Advisory Committee on Genetic Testing)[187] both of which, I think, have had quite a large influence on not just clinical genetics but on public involvement as well. In primary care we are just beginning to make progress, and I have to say that I think some of the Government initiatives that followed in the early 2000s have set us back. I think the Quality and Outcome Framework,[188] which directed primary care towards common chronic conditions, very appropriately, but also tied income generation to that, which would never be appropriate for genetics because of the ethical issues involved. Secondly, of course, and much more recently, I think the possible devolvement to 'polyclinics' – and I hate the word, but in London you're definitely going to have them – may well set back exactly what we need in genetics, where people are being looked after in a very comprehensive way by a small team of GPs and other primary care workers. All of that is meant to be possibly set aside. So I think we have got quite large difficulties. But, having said that, perhaps certainly the Genetic Education Centre in Birmingham has made enormous progress with general practice.[189] Things like the general practitioner with special interest (GPSI), which I have to say I think were a sop. Do you agree, Michael? Ten GPSIs, in this case in genetics, were never going to do anything. There were to be ten of them, and only nine were appointed.

Bobrow: So the synergies and the common interest are as obvious now as they were when I first met you, I guess. And the progress has been stodgy and it's largely organizational. And it is difficult to see how one gets through that because of the way GPs' lives are structured. That's a slightly sour note on which to end.

---

[187] The Human Genetics Commission was established in 1999, following a comprehensive review of the regulatory and advisory framework for biotechnology by the UK Government. For further details of its work, see www.hgc.gov.uk/client/Content_wide.asp?ContentId=6 (visited 21 January 2010).

[188] A voluntary annual reward and incentive programme was introduced for all GP surgeries in England by the Department of Health in 2004, detailing practice achievement results and rewarding good practice. The 2008/9 programme contained four domains: clinical care, organization, patient experience and additional services. For further details, see www.qof.ic.nhs.uk/ (visited 22 January 2010).

[189] See note 186.

Modell: We are beginning to move forward. For example, the Primary Care Genetics Society (PCGS) was established a couple of years ago (2006) to support primary care practitioners with an interest in genetic medicine.[190] The society is affiliated to the RCGP. So far it has organized two conferences focusing on education of practitioners, links with support organizations, cancer, family history, cardiovascular disease and over-the-counter testing. I believe that, so far, there are about 200 members from various health professions. This society seems to be getting off the ground rather more successfully than our previous efforts over the last two decades to raise the profile of primary care genetics.

Bobrow: I'm deeply grateful to you for a cheerier note on which to end. We are five past six, and that's past the drinking hour. I'm extremely grateful to those who set up and organized this, and to all of you for an instructive and entertaining afternoon. There's one thing that hasn't been said about the way in which clinical genetics has evolved in this country over the last little while that I wanted to add at the end. My personal experience has been that I have worked with an enormously collegiate and pleasant group of people. I'm surprised at how few people there are in the profession that I know that I've come across that I dislike – in fact, hardly one. And it's not because I'm naturally like that, believe me. I do think there's been a very good common spirit, a spirit to solve problems rather than to make them, and to work across boundaries that has been reflected in a lot of advance, and it's a great credit to all of you. So let's hope the next lot do as well. Thank you very much indeed.

Tansey: Before we all go for our glass of wine, may I say that it's been a very entertaining and fascinating afternoon. It's been a privilege to listen to some of your stories, and indeed there are at least two or three further Witness Seminars that I can see developing out of this meeting. Two of them were suggested to me at tea time. So if you do want to develop something further, please let me know, or perhaps consult Peter Harper as well, because he's now very experienced in these matters. I would like to repeat the appeal made earlier for any other archives or books, to talk to Peter Harper or Tim Powell about donating those, if you could. I would finally like to thank Peter Harper, who has worked very closely with my team in setting up this meeting, and certainly to Martin Bobrow for chairing it so ably and getting us to our drinks on time. Thank you.

---

[190] See www.pcgs.org.uk (visited 6 May 2009).

Appendix 1

Initiatives supporting clinical genetics, 1983–99

by Professor Rodney Harris[191]

Date	Initiative by	Intention	Action and publication
1983	**Clinical Genetics Society**, Working Party on the Role and Training of Clinical Geneticists, Rodney Harris (chair)		Clinical Genetics Society, Working Party (1983)
1987	**Fourth King's Fund Forum**, London, 30 November to 2 December 1987	A single body representing clinicians, genetic nurses, scientists and the public, founded by the Association of Clinical Cytogeneticists, the Clinical Genetics Society, the Clinical Molecular Geneticists and others active in this field	Harris (1988)

---

[191] Distributed at the Witness Seminar on 23 September 2008. Many initiatives were facilitated by successive Chief Medical Officers (Sir Henry Yellowlees and Sir Kenneth Calman) and by Dr Ian Lister Cheese of the DHSS.

Date	Initiative by	Intention	Action and publication
1990	**Royal College of Physicians of London**, Committee on Clinical Genetics, 1990	1. Genetic principles added to medical training 2. Manpower: consultant and senior registrar numbers and distribution: targets for regional genetic services 3. Clinical and laboratory genetic services organized 'Under one roof' 4. Establishment and training of genetic associates: key roles of nurses and others in genetic services and research 5. Special Medical Development: recognition of clinical importance of molecular genetics	1. Survey of undergraduate teaching in British medical schools reported by the medical Deans; Royal College of Physicians of London, Committee on Clinical Genetics (1990); see also Harris (1990) 2. Regular communications between RCP clinical genetics, specialty advisory committees, the JCHMT and DoH 3. Harris (1991) 4. CMO and Ministers lobbied direct 5. NHS funding for special medical development: DNA laboratories. DHSS (1987); Harris *et al.* (1989)

Date	Initiative by	Intention	Action and publication
1995	**European Commission,** Concerted action on genetics services (CAGSE)		Medical genetics recognized as a specialty in the UK in 1980; CAGSE set up a clinical geneticists' network to stimulate recognition of medical genetics across Europe; to investigate the structure, workloads and quality of genetic services and the different country response to the new opportunities and challenges of clinical genetics. Harris and Harris (1995)
1999	**Royal College of Physicians of London**, National Confidential Enquiry into counselling for genetic disorders by non-geneticists	To investigate the quality of genetic counselling provided by non-geneticist clinicians (obstetricians, paediatricians, physicians, etc.); to make general recommendations and specific standards for improving care	A review of clinical notes relating to avoidable births of Down's, neural tube defects and thalassaemia, the birth of second or subsequent CF children in family cases of multiple endocrine neoplasia type 2, to ascertain what counselling was recorded. Chief finding was extremely poor records. Subsequently a wide dissemination of reports occurred. Royal College of Physicians of London, Committee on Clinical Genetics (1999)

Appendix 2

The Association of Genetic Nurses and Counsellors (AGNC)

by Professor Heather Skirton[192]

Date	Timeline
1960s–70s	Appointment of few individual nurses and social workers to posts in genetic centres, variable roles.
1980	Genetic Nurses and Social Workers Association (GNSWA) established with nine nurses, one social worker and a written constitution.
1990	GNSWA had *c.* 100 members, one meeting a year as part of the Clinical Genetics Society (CGS) meeting. Development due to strong relationships between medical geneticists and nurses. Members had a background in midwifery and HIV; tradition of home visiting established. The Welsh Nursing Board module in genetics nursing ran for several years in the 1990s and many nurses completed this course, but most received on-the-job learning.
1992	MSc in genetic counselling at Manchester, led by Lauren Kerzin-Storrar.
1993	Education group set up by Penny Guilbert as GNSWA chair.
1995	Decision to define role and scope of practice. Study of practice led by Heather Skirton. Around 200 AGNC members.
1996	Association joined British Society for Human Genetics (BSHG) as constituent group, name changed to Association of Genetic Nurses and Counsellors (AGNC) and had a representative on BSHG council.
1997/8	Studies published in *Journal of Medical Genetics*; stand-alone Spring scientific meeting replaced CGS meeting and second meeting held with the BSHG conference.
2000	MSc in genetic counselling in Cardiff.

[192] Distributed at the Witness Seminar on 23 September 2008. For relevant publications see Skirton *et al.* (1997, 1998, 2002a, b, c, d and e, 2003); AGNC Education Working Group (2003).

Date	Timeline
2001	Launch of registration programme ; Genetic Counsellor Registration Board started, chaired by Penny Guilbert. Around 250 AGNC members. Department of Health (DoH) commissioned five reports, work led by Heather Skirton.
2002	First counsellors registered by Genetic Counsellor Registration Board.
2003–05	Input by association into DoH's *Agenda for Change* final agreement in 2004 (see www.dh.gov.uk/en/Managingyourorganisation/Workforce/Paypensionsandbenefits/Agendaforchange/dh_424). White Paper (DoH (2003)) confirmed need to grow the genetic counselling profession; new training posts funded, administered by AGNC training panel.
2008	More than 100 genetic counsellors registered, applied to Health Professions Council to register as a profession. More than 300 AGNC members.
2009	Application to Health Professions Council to register as a profession approved, as recommended by the Secretary of State for Health.

References

Adinolfi M, Alberman E. (2006) Obituary: Paul Polani. *Guardian* (17 March).

Aird I, Bentall H H, Mehigan J A, Fraser Roberts J A. (1954) The blood groups in relation to peptic ulceration and carcinoma of colon, rectum, breast and bronchus. *British Medical Journal* **ii**: 315–21.

Anon. (1963) Notes and news: medical genetics at Liverpool. *Lancet* **282**: 1074.

Anon. (1983) Editorial: the training of medical geneticists. *Lancet* **ii**: 892–3.

Anon. (1992) Editorial: screening for cystic fibrosis. *Lancet* **340**: 209–10.

Anon. (1998) Editorial: Lionel Penrose FRS, 1898–1972. *Annals of Human Genetics* **62**: 189.

Baraclough J. (1996) *One Hundred Years of Health-related Social Work, 1895–1995: Then – now – onwards.* London: British Association of Social Workers.

Baraitser M. (2004) Obituary: Robin Winter. *Guardian* (21 January). Freely available at www.guardian.co.uk/news/2004/jan/21/guardianobituaries. health (visited 17 February 2010).

Barnett R. (2004) Keywords in the history of medicine: eugenics. *Lancet* **363**: 1742.

Barr M L, Bertram E G. (1949) Letter: a morphological distinction between neurones of the male and female, and the behaviour of the nucleolar satellite during accelerated nucleoprotein synthesis. *Nature* **163**: 676–7.

Bayes T. (1763) An essay towards solving a problem in the doctrine of chances. *Philosophical Transactions [of the Royal Society]* **53**: 376–418.

Beech R, Rona R J, Mandalai S. (1994) The resource implications and service outcomes of genetic services in the context of DNA technology. *Health Policy* **26**: 171–90.

Bell E H C M. (1961) *The Story of Hospital Almoners: The birth of a profession.* London: Faber and Faber.

Bell J, Haldane J B S. (1937) The linkage between the genes for colour-blindness and haemophilia in man. *Proceedings of the Royal Society of London B* **123**: 119–50.

Blieden L C, Schneeweiss A, Neufeld H N. (1981) Primary pulmonary hypertension in leopard syndrome. *British Heart Journal* **46**: 458–60.

Bobrow M, Walker A, Walton J N. (1988) The parental origin of mutations causing Duchenne muscular dystrophy. *Archives of Neurology* **45**: 85–7.

Boon R A, Roberts D F. (1970) The social impact of haemophilia. *Journal of Biosocial Science* **2**: 237–64.

Boston S. (1981) *Will, My Son. The Life and Death of a Mongol Child.* London: Pluto Press.

Bradley D M, Parsons E P, Clarke A J. (1993) Experience with screening newborns for Duchenne muscular dystrophy in Wales. *British Medical Journal* **306**: 357–60.

British Medical Association. (1998) *Human Genetics: Choice and Responsibility.* Oxford: Oxford University Press.

Brock D J H. (1979) The First Wellcome Trust Lecture. Foeto-specific proteins in prenatal diagnosis. *Biochemical Society Transactions* **7**: 1179–95.

Brock D J H. (1990) A consortium approach to molecular genetic services. Scottish Molecular Genetics Consortium. *Journal of Medical Genetics* **27**: 8–13.

Brock D J H, Sutcliffe R G. (1972) Alpha-fetoprotein in antenatal diagnosis of anencephaly and spina bifida. *Lancet* **300**: 197–9.

Bundey S. (1996) Julia Bell MRCS LRCP FRCP (1879–1979). Steamboat lady, statistician and geneticist. *Journal of Medical Biography* **4**: 8–13.

Burn J. (1983) Clinical genetics. *British Medical Journal* **287**: 999–1000.

Burn J, Corney G. (1984) Congenital heart defects and twinning. *Acta Geneticae Medicae et Gemellologiae* **33**: 61–9.

Campbell S, Kohorn E I. (1968) Placental localization by ultrasonic compound scanning. *Journal of Obstetrics and Gynaecology of the British Commonwealth* **75**: 1007–13.

Carter C O. (1956) Letter: intelligence and fertility. *British Medical Journal* **i**: 47–8.

Carter C O. (1966) Differential fertility by intelligence. In Meade J E, Parkes A S (eds) *Genetic and Environmental Factors in Human Ability: A symposium held by the Eugenics Society in September–October 1965.* Edinburgh and London: Oliver and Boyd: 185–200.

Carter C O, Roberts J A, Evans K A, Buck A R. (1971) Genetic clinic: a follow-up. *Lancet* **297**: 281–5.

Cheese I A F L. (1997) Contribution to a meeting held on Thursday 23 October 1997 at Manchester Royal Infirmary, to celebrate the achievements of Professor Rodney Harris CBE MD FRCP FRCPath, professor of medical genetics in the University of Manchester. Unpublished manuscript, deposited in archives and manuscripts, Wellcome Library, London, at GC/253.

Christie D A, Tansey E M. (eds) (2003) *Genetic Testing.* Wellcome Witnesses to Twentieth Century Medicine, vol. 17. London: Wellcome Trust Centre for the History of Medicine at UCL. Freely available online at www.ucl.ac.uk/histmed/publications/wellcome_witnesses_c20th_med

Christie D A, Tansey E M. (eds) (2006) *Development of Physics Applied to Medicine in the UK, 1945–90.* Wellcome Witnesses to Twentieth Century Medicine, vol. 28. London: Wellcome Trust Centre for the History of Medicine at UCL. Freely available online at www.ucl.ac.uk/histmed/publications/wellcome_witnesses_c20th_med

Clark R W. (1968) *JBS: The Life and Work of J B S Haldane.* London: Hodder and Stoughton.

Clarke A J. (1994a) The genetic testing of children: Working Party of the Clinical Genetics Society (UK). *Journal of Medical Genetics* **31**: 785–97.

Clarke A J. (ed.) (1994b) *Genetic Counselling: Practice and principles.* London: Routledge.

Clarke A J. (1995) The genetic testing of children. *Journal of Medical Genetics* **32**: 492.

Clarke A J. (1997a) Prenatal genetic screening. In Harper P S, Clarke A. (eds) *Genetics, Society and Clinical Practice.* Oxford: Bios Scientific Publishers: 119–40.

Clarke A J. (1997b) Challenges to genetic privacy. In Harper P S, Clarke A. (eds) *Genetics, Society and Clinical Practice.* Oxford: Bios Scientific Publishers: 149–64.

Clarke A J. (1997c) Parents' responses to predictive genetic testing in their children. *Journal of Medical Genetics* **34**: 174–5.

Clarke A J, Harper P S. (1992) Genetic testing for hypertrophic cardiomyopathy. *New England Journal of Medicine* **327**: 1175–6.

Clarke C A. (1974) Lionel Penrose: some aspects of his life and work. *Journal of the Royal College of Physicians of London* **8**: 237–50.

Clarke C A. (1985) Robert Russell Race. *Biographical Memoirs of Fellows of the Royal Society* **31**: 455–92.

Clarke C A, Sheppard P M. (1965) Letter: prevention of rhesus haemolytic disease. *Lancet* **ii**: 343.

Clarke C A, Edwards J W, Haddock D R W, Howel Evans A W, McConnell R B, Sheppard P M. (1956) ABO blood groups and secretor character in duodenal ulcer. *British Medical Journal* **ii**: 725–31.

Clegg J B, Weatherall D J, Na-Nakorn S, Wasi P. (1968) Haemoglobin synthesis in β-thalassaemia. *Nature* **220**: 664–8.

Clinical Genetics Society, Working Party. (1983) Report of the Working Party on the Role and Training of Clinical Geneticists. *Bulletin of the Eugenics Society* **5** (Suppl.): 1–30.

Consultant Paediatricians and Medical Officers of Health, South-East Scotland Hospital Region. (1968) Report on population screening by Guthrie test for phenylketonuria in South-East Scotland. *British Medical Journal* **i**: 674–6.

Craufurd D, Harris R. (1989) Predictive testing for Huntington's disease. *British Medical Journal* **298**: 892.

Craufurd D, Donnai D, Kerzin-Storrar L, Osborn M. (1990) Testing of children for 'adult' genetic diseases. *Lancet* **335**: 1406.

Crosse V M, Corney G. (1961) The use of adrenocorticotrophic hormone in neonatal jaundice in premature babies. *Proceedings of the Royal Society of Medicine* **54**: 737–9.

Cudworth A G, Woodrow J C. (1974) Letter: HL-A antigens and diabetes mellitus. *Lancet* **ii**: 1153.

Davidson W M, Smith D R. (1954) A morphological sex difference in the polymorphonuclear neutrophil leucocytes. *British Medical Journal* **ii**: 6–7.

Davies S J, Farndon P, Harper P S. (1998) *Commissioning Clinical Genetic Services: A report.* London: Royal College of Physicians.

Dean M. (2010) Obituary: Sir Donald Acheson. *Guardian* (16 January): 43.

Dennis N R, Evans K, Clayton B, Carter C O. (1976) Use of creatine kinase for detecting severe X-linked muscular dystrophy carriers. *British Medical Journal* **ii**: 577–9.

Department of Health. (1989) *On the State of the Public Health: The annual report of the Chief Medical Officer of the Department of Health for the year 1988.* London: HMSO.

Department of Health. (1992a) *On the State of the Public Health: The annual report of the Chief Medical Officer of the Department of Health for the year 1991.* London: HMSO.

Department of Health. (1992b) *Report of the Committee on the Ethics of Gene Therapy,* Cmnd 1788. London: HMSO.

Department of Health. (1992c) *Government Response to the Fourth Report from the Health Committee, Session 1990/1. Preconception care.* London: HMSO.

Department of Health. (1993) *On the State of the Public Health: The annual report of the Chief Medical Officer of the Department of Health for the year 1992.* London: HMSO.

Department of Health, CMO and CNO. (1993) *Services for Genetic Disorders,* Professional letter, PL/CMO(93)5 and PL/CNO(93)4 with enclosure entitled *Guide: Population needs and genetic services.* London: HMSO.

Department of Health. (2000) *The NHS Plan: A plan for investment, a plan for reform,* Cmnd 4818-I. London: HMSO.

Department of Health. (2003) *Our Inheritance, Our Future: Realising the potential of genetics in the NHS,* 2 vols, Cmnd 5791–I&II. London: HMSO. Freely available at: www.dh.gov.uk/dr_consum_dh/groups/dh_digitalassets/ @dh/@en/documents/digitalasset/dh_4019239.pdf (visited 22 January 2010).

Department of Health and Social Security. (1987) *Special Medical Development in Clinical Genetics. Interim Report: Clinical effectiveness in the service context.* Medical Division, DHSS, unpublished mimeo (lodged at copyright libraries).

Department of Health and Social Security, Scottish Home and Health Department, Welsh Office, Standing Advisory Committee. (1972) *Human Genetics.* London: HMSO.

Doll R, Esterman A, Gilliland I, Joules H, Leys D, Penrose L, Pollock M. (1951) Letter: prospect of war. *Lancet* **257**: 415.

Donnai D. (n.d., *c.* 2004) Professor Robin Michael Winter, 1950–2004: An Appreciation. Freely available at http://www.nesc.ac.uk/esi/events/293/AJHG_on_Robin_Winter.pdf (visited 19 January 2010).

Elles R. (ed.) *Molecular Diagnosis of Genetic Diseases.* Totowa, NJ: Humana Press.

Emery A E H. (1968a) *Inaugural Lecture 35: Genetics in Medicine,* delivered on 29 April 1968. Edinburgh: University of Edinburgh.

Emery A E H. (1968b) *Elements of Medical Genetics.* Edinburgh; London: E & S Livingstone. (Originally published as *Heredity, Disease and Man.* Berkeley, CA: University of California Press, 1968).

Emery A E H, Pullen I M. (eds) (1984) *Psychological Aspects of Genetic Counselling.* London: Academic Press.

Emery A E H, Brough C, Craufurd M, Harper P, Harris R, Oakshott G. (1978) A report on genetic registers, based on the report of the Clinical Genetics Society Working Party. *Journal of Medical Genetics* **15**: 435–42.

Featherstone K, Atkinson P A, Bharadwaj A, Clarke A J. (2006) *Risky Relations. Family and kinship in the era of new genetics.* Oxford: Berg Publishers.

Ferguson-Smith M A. (2008) Cytogenetics and the evolution of medical genetics. *Genetics in Medicine* **10**: 553–9.

Fine B, Baker D, Fiddler M and ABGC Consensus Development Consortium. (1996) Practice-based competencies for accreditation of and training in graduate programs in genetic counseling. *Genetic Counseling* **5**: 113–21.

Finn R, Clarke C A, Donohoe W T, McConnell R B, Sheppard P M, Lehane D, Kulke W. (1961) Experimental studies on the prevention of Rh haemolytic disease. *British Medical Journal* **i**: 1486–90.

Ford C E, Jones K W, Miller O J, Mittwoch U, Penrose L S, Ridler M, Shapiro A. (1959a) The chromosomes in a patient showing both mongolism and the Klinefelter syndrome. *Lancet* **i**: 709–10.

Ford C E, Jones K W, Polani P E, de Almeida J C, Briggs J H. (1959b) A sex-chromosome anomaly in a case of gonadal dysgenesis (Turner's syndrome). *Lancet* **i**: 711–13.

Forrest K, Simpson S A, Wilson B J, van Teijlingen E R, McKee L, Haites N, Matthews E. (2003) To tell or not to tell: barriers and facilitators in family communication about genetic risk. *Clinical Genetics* **64**: 317–26.

Fraser Roberts J A. (1940) *An Introduction to Medical Genetics.* London: Oxford University Press.

Fuhrmann W, Vogel F. (1969) *Genetic Counseling: A guide for the practicing physician*, Kurth S (trns), Heidelberg science library vol. 10. New York, NY: Springer-Verlag New York Inc.

Gale E A G. (2001) The discovery of type 1 diabetes. *Diabetes* **50**: 217–26.

Geary J, Sasieni P, Houlston R, Izatt L, Eeles R, Payne S J, Fisher S, Hodgson S V. (2008) Gene-related cancer spectrum in families with hereditary non-polyposis colorectal cancer (HNPCC). *Familial Cancer* **7**: 163–72.

Gillam S J, Macdonald I. (eds) (2001) Alan Carruth Stevenson. *Munk's Roll* **10**: 469–70.

Griffiths S. (2006) The role of the postgraduate medical education and training board. *Archives of Disease in Childhood* **91**: 195–7.

Gunther M, Penrose L S. (1935) The genetics of epiloia. *Journal of Genetics* **35**: 413–30.

Hagard S, Carter F A. (1976) Preventing the birth of infants with Down's syndrome: a cost–benefit analysis. *British Medical Journal* **i**: 53–6.

Hamerton J L, Giannelli F, Polani P E. (1965) Cytogenetics of Down's syndrome (mongolism). I. Data on a consecutive series of patients referred for genetic counselling and diagnosis. *Cytogenetics* **4**: 171–85.

Harkness R A, Cockbury F. (eds) *The Cultured Cell and Inherited Metabolic Disease.* Lancaster: MTP Press Ltd.

Harnden D G. (1960) A human skin culture technique used for cytological examinations. *British Journal of Experimental Pathology* **41**: 31–7.

Harnden D G. (1979) Cell biology and cell culture methods: a review. In Harkness R A, Cockbury F. (eds) *The Cultured Cell and Inherited Metabolic Disease.* Lancaster: MTP Press Ltd: 3–15.

Harnden D G. (1996) Early studies on human chromosomes. *BioEssays* **18**: 163–8.

Harnden D G, Miller O J, Penrose L S. (1960) The Klinefelter mongolism type of double aneuploidy. *Annals of Human Genetics* **24**: 165–9.

Harper P S. (1981) *Practical Genetic Counselling.* Bristol: Wright.

Harper P S. (1985) Editorial: Sir Cyril Clarke and the *Journal of Medical Genetics*: an appreciation. *Journal of Medical Genetics* **22**: 417.

Harper P S. (1995) Genetic testing, common diseases and health service provision. *Lancet* **346**: 1645–6.

Harper P S. (1997) Presymptomatic testing for late-onset genetic disorders: lessons from Huntington's Disease. In Harper P S, Clarke A. (eds) *Genetics, Society and Clinical Practice.* Oxford: Bios Scientific Publishers: 31–44.

Harper P S. (2004) *Practical Genetic Counselling*, 6th edn. London: Arnold.

Harper P S. (2005) Julia Bell and the *Treasury of Human Inheritance. Human Genetics* **116**: 422–32.

Harper P S. (2006) *First Years of Human Chromosomes.* Oxford: Scion.

Harper P S. (2007) Paul Polani and the development of medical genetics. *Human Genetics* **120**: 723–31.

Harper P S. (2008) *A Short History of Medical Genetics.* New York, NY: Oxford University Press.

Harper P S, Clarke A J. (1990) Should we test children for 'adult' genetic diseases? *Lancet* **335**: 1205–6.

Harper P S, Clarke A. (eds) (1997) *Genetics, Society and Clinical Practice.* Oxford: Bios Scientific Publishers.

Harper P S, Pierce K. (2010) The Human Genetics Historical Library: an international resource for geneticists and historians. *Clinical Genetics* **77**: 214–20.

Harper P S, Hughes H E, Raeburn J A. (1996) *Clinical Genetics Services into the Twenty-first Century: A report.* London: Royal College of Physicians.

Harper P S, Tyler A, Smith S, Jones P, Newcombe R G, McBroom V. (1981) Decline in the predicted incidence of Huntington's chorea associated with systematic genetic counselling and family support. *Lancet* **ii**: 411–13.

Harris H. (1959) *Human Biochemical Genetics.* London: Cambridge University Press.

Harris H. (1970) Genetical theory and the 'inborn errors of metabolism'. *British Medical Journal* **i**: 321–7.

Harris H. (1973) Lionel Sharples Penrose (1898–1972). *Biographical Memoirs of Fellows of the Royal Society* **19**: 521–61.

Harris H. (1974) The development of Penrose's ideas in genetics and psychiatry. *British Journal of Psychiatry* **125**: 529–36.

Harris H, Scotcher D, Hartley N, Wallace A, Craufurd D, Harris R. (1993) Cystic fibrosis carrier testing in early pregnancy by general practitioners. *British Medical Journal* **306**: 1580–3.

Harris R. (1988) King's Fund forum, screening for fetal and genetic abnormality: the need for a British Association for Genetic Medicine. *Journal of Medical Genetics* **25**: 145–6.

Harris R. (1990) Letter: physicians and other non-geneticists strongly favour teaching genetics to medical students in the UK. *American Journal of Human Genetics* **47**: 750–2.

Harris R. (1991) *Clinical Genetic Services in 1990 and Beyond: A report of the Clinical Genetics Committee of the Royal College of Physicians.* London: Royal College of Physicians of London.

Harris R. (ed.) (1997) Genetic services in Europe: a comparative study of 31 countries by the concerted action on genetic services in Europe. *European Journal of Medical Genetics* **5**(suppl.): 1–220.

Harris R, Harris H J. (1995) Primary care for patients at genetic risk: a priority for EC concerted action on genetics services. *British Medical Journal* **311**: 579–80.

Harris R, Harris H J. (1999) Clinical governance and genetic medicine: specialist genetic centres and the Confidential Enquiry into Counselling for Genetic Disorders by non-geneticists (CEGEN). *Journal of Medical Genetics* **36**: 350–1.

Harris R, Timson J, Wentzel J. (1969) The HLA leucocyte system in Down syndrome. *Transplantation Proceedings* **1**: 122–3.

Harris R, Lane B, Harris H, Williamson P, Dodge J, Modell B, Ponder B, Rodeck C, Alberman E. (1999) National Confidential Enquiry into counselling for genetic disorders by non-geneticists: general recommendations and specific standards for improving care. *British Journal of Obstetrics and Gynaecology* **16**: 658–63.

Harris R, Elles R, Craufurd D, Dodge A, Ivinson A, Hodgkinson K, Mountford R, Schwartz M, Strachan T, Read A. (1989) Molecular genetics in the National Health Service in Britain. *Journal of Medical Genetics* **26**: 219–25.

Hibbard B M, Roberts C J, Elder G H, Evans K T, Laurence K M. (1985) Can we afford screening for neural tube defects? The South Wales experience. *British Medical Journal* **290**: 293–5.

Hill A B. (1937) *Principles of Medical Statistics.* London: The Lancet Ltd.

Hill A B. (1988) Obituary: Dr J A Fraser Roberts. *Journal of the Royal Statistical Society Series A (Statistics in Society)* **151**: 359–60.

Hodgson S V. (1999) Cancer genetics services in the UK. *Disease Markers* **15**: 44–5.

Hodgson S V, Walker A, Cole C, Hart K, Johnson L, Heckmatt J, Dubowitz V, Bobrow M. (1987) The application of linkage analysis to genetic counselling in families with Duchenne or Becker muscular dystrophy. *Journal of Medical Genetics* **24**: 152–9.

Hodgson S V, Milner B, Brown I, Bevilacqua G, Chang-Claude J, Eccles D, Evans G, Gregory H, Møller P, Morrison P, Steel M, Stoppa-Lyonnet D, Vasen H, Haites N. (1999) Cancer genetics services in Europe. *Disease Markers* **15**: 3–13.

Hopkinson D. (1994) Professor Harry Harris. *Independent* (22 July).

Houlston R S, Collins A, Slack J, Campbell S, Collins W P, Whitehead M I, Morton N E. (1991) Genetic epidemiology of ovarian cancer: segregation analysis. *Annals of Human Genetics* **55**: 291–9.

Houlston R S, Bourne T H, Davies A, Whitehead M I, Campbell S, Collins W P, Slack J. (1992) Use of family history in a screening clinic for familial ovarian cancer. *Gynecologic Oncology* **47**: 247–52.

House of Commons, Health Committee. (1991). *Maternity Services: Preconception care: Fourth report from the Health Committee*, HC430-I&II, Session 1990/1, 2 vols. London: HMSO.

House of Commons, Science and Technology Committee. (1995) *Human Genetics: The science and its consequences: Third Report, Vol.1: Report and minutes of proceedings*, HC41-I, Session 1994/5. London: HMSO.

House of Commons, Science and Technology Committee. (1996) *Human Genetics: The Government's Response to the Third Report from the Science and Technology Committee*, HC231-I, Session 1994/5. London: HMSO.

House of Lords, Science and Technology Committee. (2009) *Second Report: Genomic Medicine*, HL107-I&II, Session 2008/9, 2 vols. London: The Stationery Office. Browsable copy at www.publications.parliament.uk/pa/ld200809/ldselect/ldsctech/107/10702.htm (visited 5 March 2010).

Howel-Evans W, McConnell R B, Clarke C A, Sheppard P M. (1958) Carcinoma of the oesophagus with keratosis palmaris et plantaris (tylosis): a study of two families. *Quarterly Journal of Medicine* **27**: 413–29.

Hubbard R. (1997) Abortion and disablity: who should and who should not inhabit the world? In Davis L J. (ed.) *The Disability Studies Reader*. London; New York, NY: Routledge: 187–200.

Human Genetics Commission. (2006) *Making Babies: Reproductive decisions and genetic technologies*. London: Human Genetics Commission. Freely available at www.hgc.gov.uk/UploadDocs/DocPub/Document/Making%20Babies%20Report%20-%20final%20pdf.pdf (visited 16 April 2010).

Hutton K. (1953) Intelligence quotients and differential fertility – some observations from Winchester College. *Eugenics Review* **44**: 205–15.

Iselius L, Slack J, Littler M, Morton N E. (1991) Genetic epidemiology of breast cancer in Britain. *Annals of Human Genetics* **55**: 151–9.

Jacobs P A, Strong J A. (1959) A case of human intersexuality having a possible XXY sex-determining mechanism. *Nature* **183**: 302–3.

Jacobs P A, Baikie A G, Court Brown W M, Strong J A. (1959) The somatic chromosomes in mongolism. *Lancet* **i**: 710.

Johnston A W. (1978) Training of medical geneticists. *Journal of Medical Genetics* **15**: 260–1.

Johnston A W. (1979) The training of medical geneticists in Britain. *American Journal of Medical Genetics* **3**: 7–9.

Kelly K F. (2002) The Scottish molecular genetics consortium: 15 years on. *Health Bulletin* **60**: 83–90.

Kerr C B. (1968) X-linked haematological traits. *Bibliotheca Haematologica* **29**: 59–70.

Kerzin-Storrar L. (1996) Genetic counselling and molecular testing. In Elles R. (ed.) *Molecular Diagnosis of Genetic Diseases.* Totowa, NJ: Humana Press: 205–17. At http://springerprotocols.com/Abstract/doi/10.1385/0-89603-346-5:205 (visited 9 February 2010).

Kettlewell H B D. (1956) *Further Selection Experiments on Industrial Melanism in the Lepidoptera.* London: Oliver & Boyd.

Kevles D J. (1995) *In the Name of Eugenics: Genetics and the use of human heredity.* Cambridge, MA: Harvard University Press.

Kevles D J. (2007) Lionel Penrose, mental deficiency and human genetics. In Mayo O, Leach C. (eds) *Fifty Years of Human Genetics: A* Festschrift *and* liber amicorum *to celebrate the life and work of George Robert Fraser.* Kent Town, South Australia: Wakefield Press: 39–47.

Lanchbury J S, Papiha S S, Roberts D F. (1990) Genetic variation in north-east England. *Annals of Human Biology* **17**: 265–76.

Lane B, Challen K, Harris H J, Harris R. (2001) Existence and quality of written antenatal screening policies in the UK: postal survey. *British Medical Journal* **322**: 22–3.

Lassonde M, Trudeau J G, Girard C. (1970) Generalized lentigines associated with multiple congenital defects (leopard syndrome). *Canadian Medical Association Journal* **103**: 293–4.

Laxova R. (1998) Lionel Sharples Penrose, 1898–1972: A personal memoir in celebration of the centenary of his birth. *Genetics* **150**: 1333–40.

Leeming W. (2010) Tracing the shifting sands of 'medical genetics': what's in a name? *Studies in History and Philosophy of Biological and Biomedical Sciences* **41**: 50–60.

Lejeune J, Turpin R, Gautier M. (1959) Les chromosomes humains en culture de tissues [Human chromosomes in tissue cultures]. *Comptes Rendus Hebdomadaires des Séances de l'Académie des Sciences* **248**: 602–3.

Li C C. (2000) Progressing from eugenics to human genetics. *Human Heredity* **50**: 22–33.

Lucas M, Mullarkey M, Abdulla U. (1972) Study of chromosomes in the newborn after ultrasonic fetal heart monitoring in labour. *British Medical Journal* **iii**: 795–6.

Lyon M F. (2001) Charles Edmund Ford FRS. *Biographical Memoirs of Fellows of the Royal Society* **47**: 189–202.

MacGillivray I. (2002) Gerald Corney, 1928–2001. *Twin Research* **5**: 65.

Macklin M T. (1933) Medical genetics: an essential part of the medical curriculum from the standpoint of prevention. *Journal of the Association of American Medical Colleges* **8**: 291–301.

Mayo O, Leach C. (eds) (2007) *Fifty Years of Human Genetics: A* Festschrift *and* liber amicorum *to celebrate the life and work of George Robert Fraser.* Kent Town, South Australia: Wakefield Press.

McKusick V A. (2001) Persisting memories of Cyril Clarke in Baltimore. *Journal of Medical Genetics* **38**: 284 (doi:10.1136/jmg.38.5.284).

Meade J E, Parkes A S. (eds) (1966) *Genetic and Environmental Factors in Human Ability: A symposium held by the Eugenics Society in September–October 1965.* Edinburgh and London: Oliver and Boyd.

Medical Research Council, Working Party on Amniocentesis (1978) An assessment of the hazards of amniocentesis. *British Journal of Obstetrics and Gynaecology* **85**(suppl.): 1–41.

Medical Research Council, Department of Health and Social Security, joint working group. (1985) *Report on Genetic Counselling and Service Implications of Clinical Genetics Research.* MRC 79/588. Available at copyright libraries.

Mittwoch U. (1958a) The polymorphonuclear lobe count in mongolism and its relation to the total leucocyte count. *Journal of Mental Deficiency Research* **2**: 75–80.

Mittwoch U. (1958b) The leucocyte count in children with mongolism. *Journal of Mental Science* **104**: 457–60.

Mittwoch U. (1995) Living history – biography. *American Journal of Medical Genetics* **55**: 3–11.

Modell B. (1997) Delivering genetic screening to the community. *Annals of Medicine* **29**: 591–9.

Modell M, Wonke B, Anionwu E, Khan M, Tai S S, Lloyd M, Modell B. (1998) A multidisciplinary approach for improving services in primary care: randomised controlled trial of screening for haemoglobin disorders. *British Medical Journal* **317**: 788–91.

Mouzouras M, Camba L, Ioannou P, Modell B, Constantinides P, Gale R. (1980) Thalassaemia as a model of recessive genetic disease in the community. *Lancet* **ii**: 574–8.

Murday V, Slack J. (1989) Inherited disorders associated with colorectal cancer. *Cancer Surveys* **8**: 139–57.

Murphy E A, Mutalik G S. (1969). The application of Bayesian methods in genetic counselling. *Human Heredity* **19**: 126–51.

Mutton D, Alberman E, Hook E B for the National Down Syndrome Cytogenetic Register and the Association of Clinical Cytogeneticists. (1996) Cytogenetic and epidemiological findings in Down syndrome, England and Wales, 1989 to 1993. *Journal of Medical Genetics* **33**: 387–94.

Nance W. (2004) Obituary: Robin M Winter. *Journal of Medical Genetics* **41**: 718.

National Medical Consultative Committee, Working Group. (1986) *Clinical Genetic Services in Scotland.* Edinburgh: Scottish Home and Health Department.

Ness A R, Reynolds L A, Tansey E M. (eds) (2002) *Population-based Research in South Wales: The MRC Pneumoconiosis Research Unit and the MRC Epidemiology Unit.* Wellcome Witnesses to Twentieth Century Medicine, vol. 13. London: The Wellcome Trust Centre for the History of Medicine at UCL. Freely available online at www.ucl.ac.uk/histmed/publications/wellcome_witnesses_c20th_med

Norman A P. (1983) Sir Alan Aird Moncrieff. *Munk's Roll* 7: 343–6.

Nuffield Council on Bioethics. (1993) *Genetic Screening – Ethical Issues.* London: Nuffield Council on Bioethics.

Osler W. (1904) *Aequanimitas: With other addresses to medical students, nurses and practitioners of medicine.* London: H K Lewis.

Ottolenghi S, Lanyon W G, Paul J, Williamson R, Weatherall D J, Clegg J B, Pritchard J, Pootrakul S, Boon W H. (1974) The severe form of α-thalassaemia is caused by a haemoglobin gene deletion. *Nature* **251**: 389–92.

Papiha S S, Roberts D F. (1972) Adenosine deaminase (ADA) polymorphism in Northern England. *Humangenetik* **15**: 279–81.

Parens E, Asch A. (2000) The disability rights critique of prenatal genetic testing. In Parens E, Asch A. (eds) *Prenatal Testing and Disability Rights.* Washington, DC: Georgetown University Press: 3–43.

Pearn J H. (1973a) Patients' subjective interpretation of risks offered in genetic counselling. *Journal of Medical Genetics* **10**: 129–34.

Pearn J H. (1973b) The gene frequency of acute Werdnig-Hoffmann disease (SMA-type-1): a total population survey in North-east England. *Journal of Medical Genetics* **10**: 260–5.

Pearn J H, Gardner-Medwin D, Wilson J. (1978) A clinical study of chronic childhood spinal muscular atrophy. A review of 141 cases. *Journal of the Neurological Sciences* **38**: 23–37.

Pembrey M E. (1987) Obituary: Dr John Alexander Fraser Roberts. *Journal of Medical Genetics* **24**: 442–4.

Pembrey M E. (1994) *Genetics: A simple guide.* London: Channel Four Television in association with the Progress Educational Trust.

Pembrey M E; ALSPAC Study Team. (2004) The Avon Longitudinal Study of Parents and Children (ALSPAC): a resource for genetic epidemiology. *European Journal of Endocrinology* 151 (Suppl. 3): U125-9.

Penrose L S. (1938) *A Clinical and Genetic study of 1280 Cases of Mental Defect,* Medical Research Council Special Report no. 229 (the Colchester survey). London: HMSO for the Privy Council of the MRC. Reprinted in 1975 by the Institute for Research into Mental and Multiple Handicap.

Penrose L S. (1946) Phenylketonuria: a problem in eugenics. *Lancet* **i**: 949–53.

Penrose L S. (1952) *On the Objective Study of Crowd Behaviour.* London: H K Lewis.

Penrose L S. (1972) *The Biology of Mental Defect,* 4th rev. edn. London: Sidgwick and Jackson.

Penrose L S. (1998) Phenylketonuria: a problem in eugenics. *Annals of Human Genetics* **62**: 193–202.

Penrose L S, Delhanty J D. (1961) Familial Langdon Down anomaly with chromosomal fusion. *Annals of Human Genetics* **25**: 243–52.

Penrose L S, Ellis J R, Delhanty J D. (1960) Chromosomal translocations in mongolism and in normal relatives. *Lancet* **276**: 409–10.

Penrose O. (1998) Lionel Penrose FRS, human geneticist and human being. In Povey S, Press M. (eds) *Pioneer in Human Genetics. Report on a symposium to celebrate the centenary of the birth of Lionel Penrose, held on 12–13 March 1998,* produced by Sue Povey and Marina Press. London: Centre for Human Genetics at UCL: 8–14.

Piggott M, Wilkinson P, Bennett J. (1994) Implementation of an antenatal serum screening programme for Down's syndrome in two districts (Brighton and Eastbourne). The Brighton and Eastbourne Down's Syndrome Screening Group. *Journal of Medical Screening* **1**: 45–9.

Pirie N W. (1966) John Burdon Sanderson Haldane. *Biographical Memoirs of Fellows of the Royal Society* **12**: 219–50.

Platt R. (1963) *Doctor and Patient: Ethics, morale, government,* Rock Carling fellowship 1963. London: The Nuffield Provincial Hospitals Trust.

Platt R. (1972) *Private and Controversial.* London: Cassell.

Polani P E. (1987) John Alexander Fraser Roberts. *Biographical Memoirs of Fellows of the Royal Society* **38**: 307–22.

Polani P E, Hunter W F, Lennox B. (1954) Chromosomal sex in Turner's syndrome with coarctation of the aorta. *Lancet* **267**: 120–1.

Polani P E, Lessof M H, Bishop P M F. (1956) Colour-blindness in ovarian agenesis (gonadal dysplasia). *Lancet* **271**: 118–20.

Polani P E, Briggs J H, Ford C E, Clarke C M, Berg J M. (1960) A mongol girl with 46 chromosomes. *Lancet* **i**: 721–4.

Polani P E, Alberman E, Alexander B J, Benson P F, Berry A C, Blunt S, Daker M G, Fensom A H, Garrett D M, McGuire V M, Roberts J A, Seller M J, Singer J D. (1979) Sixteen years' experience of counselling, diagnosis and prenatal detection in one genetic centre: Progress, results, and problems. *Journal of Medical Genetics* **16**: 166–75.

Polkinghorne J. (1989) *Review of the Guidance on the Research Use of Fetuses and Fetal Material*, Cmnd 762. London: HMSO.

Ponder B A. (1987) Familial cancer: opportunities for clinical practice and research. *European Journal of Surgical Oncology* **13**: 463–73.

Ponder B A. (1991) Genetic predisposition to cancer. *British Journal of Cancer* **64**: 203–4.

Povey S, Press M. (eds) (1998) *Penrose: Pioneer in human genetics: Report on a symposium held to celebrate the centenary of the birth of Lionel Penrose*, held 12th and 13th March 1998, produced by Sue Povey and Marina Press. London: Centre for Human Genetics at UCL.

Price Evans D A. (1962) Pharmacogenetics. *Acta Geneticae Medicae et Gemellologiae* **11**: 338–50.

Rafi I, Qureshi N, Lucassen A, Modell M, Elmslie F, Kai J, Kirk M, Starey N, Goff S, Brennan P, Hodgson S. (2009) 'Over-the-counter' genetic testing: what does it really mean for primary care? *British Journal of General Practice* **59**: 283–7.

Read A P, Donnai D. (2007) *New Clinical Genetics.* Bloxham: Scion.

Reed S C. (1955) *Counseling in Medical Genetics.* Philadelphia, PA; London: W B Saunders.

Reed S C. (2005) A short history of human genetics in the USA. *American Journal of Medical Genetics* **3**: 282–95.

Reynolds L A, Tansey E M. (eds) (2006) *Cholesterol, Atherosclerosis and Coronary Disease in the UK, 1950–2000.* Wellcome Witnesses to Twentieth Century Medicine, vol. 27. London: Wellcome Trust Centre for the History of Medicine at UCL. Freely available online at www.ucl.ac.uk/histmed/ publications/wellcome_witnesses_c20th_med

Reynolds L A, Tansey E M. (eds) (2007) *Medical Ethics Education in Britain, 1963–93.* Wellcome Witnesses to Twentieth Century Medicine, volume 31. London: The Wellcome Trust Centre for the History of Medicine at UCL. Freely available online at www.ucl.ac.uk/histmed/publications/wellcome_ witnesses_c20th_med

Richmond C. (2006) Professor Paul Polani: geneticist in paediatric research. *Independent* (21 March).

Roberts D F. (1968) Genetic effects of population size reduction. *Nature* **220**: 1084–8.

Robson B. (1996) Obituary: Sylvia Lawler. *British Medical Journal* **312**: 906.

Rocker I, Laurence K M. (eds) (1981) *Fetoscopy.* Amsterdam; Oxford: Elsevier/ North-Holland Biomedical Press.

Rose F C. (1992) Lord Walton of Detchant. *Postgraduate Medical Journal* **68**: 497–9.

Rosser E M, Wilson L M. (2004) Robin M Winter – a colleagues' perspective. *Journal of Medical Genetics* **41**: 718–20.

Royal College of Physicians, Clinical Genetics Committee, Working Party. (1990) Teaching genetics to medical students: report of a working party of the clinical genetics committee of the Royal College of Physicians. *Journal of the Royal College of Physicians of London* **24**: 80–4.

Royal College of Physicians, Clinical Genetics Committee. (1991a) *Retention of Medical Records in relation to Genetic Diseases: A report.* London: Royal College of Physicians. First edition prepared by Dr Sarah Bundey, second edition 1998.

Royal College of Physicians, Committees on Ethical Issues in Medicine and Clinical Genetics. (1991b) *Ethical Issues in Clinical Genetics,* prepared by Janet Radcliffe Richards with Martin Bobrow. London: Royal College of Physicians of London.

Santesson B, Åkesson H-O, Böök J A, Brosset A. (1969) Letter: karyotyping human amniotic-fluid cells. *Lancet* **294**: 1067–8.

Scrimgeour J B. (1978) Antenatal diagnosis in early pregnancy. *British Journal of Hospital Medicine* **19**: 565–73.

Shackley P, McGuire A, Boyd P A, Dennis J, Fitchett M, Kay J, Roche M, Wood P. (1993) An economic appraisal of alternative pre-natal screening programmes for Down's syndrome. *Journal of Public Health Medicine* **15**: 175–84.

Shamsadini S, Abazardi H, Shamsadini F. (1999) Leopard syndrome. *Lancet* **354**: 1530.

Shotter E. (2004) Professor the Reverend Canon G R Dunstan. *Journal of Medical Ethics* **30**: 233–4.

Silverman M E. (2003) Obituary: Maurice Campbell: first editor of *Heart. Heart* **89**: 1379–81.

Skirton H, Barnes C, Curtis G, Walford-Moore J. (1997) The role and practice of the genetic nurse. *Journal of Medical Genetics* **34**: 141–7.

Skirton H, Barnes C, Guilbert P, Kershaw A, Kerzin-Storrar L, Patch C, Curtis G, Walford-Moore J. (1998) Recommendations for education and training of genetic nurses and counsellors in the UK. *Journal of Medical Genetics* **35**: 410–12.

Skirton H, Kerzin-Storrar L, Patch C for AGNC Education Working Group. (2002a) Reports for the Genetics Policy Unit, Department of Health, UK. *Report A. A review of studies related to the provision and outcomes of genetic services.* Unpublished.

Skirton H for AGNC Education Working Group, convenor Dr Heather Skirton. (2002b) Reports for the Genetics Policy Unit, Department of Health, UK. *Report B. An interim report on the Association of Genetic Nurses and Counsellors Registration process.* Unpublished.

Skirton H, Barnes C, Patch C for AGNC Education Working Group. (2002c) Reports for the Genetics Policy Unit, Department of Health, UK. *Report C. Report on the professional background, working contracts and training needs of genetic counsellors in the United Kingdom and Ireland.* Unpublished.

Skirton H, Barnes C, Kerzin-Storrar L for AGNC Education Working Group. (2002d) Reports for the Genetics Policy Unit, Department of Health, UK. *Report D. Expanding training capacity for genetic counsellors: proposals for creation of training posts and clinical placement centres.* Unpublished.

Skirton H, Barnes C, Dolling C, Guilbert P, Kershaw A, Kerzin-Storrar L, Patch C, Stirling D for AGNC Education Working Group. (2002e) *Report E. Education in genetics for health professionals not working in specialist genetic units*, Reports for the Genetics Policy Unit, Department of Health, UK. Unpublished.

Skirton H, Kerzin-Storrar L, Patch C, Barnes C, Guilbert P, Dolling C, Kershaw A, Baines E, Stirling D. (2003) Genetic counsellors – a registration system to assure competence in practice in the UK. *Community Genetics* **6**:182–3.

Slack J. (1969) Risks of ischaemic heart disease in familial hyperlipoproteinaemic states. *Lancet* **ii**: 1380–2.

Slack J, Evans K A. (1966) The increased risk of death from ischaemic heart disease in first degree relatives of 121 men and 96 women with ischaemic heart disease. *Journal of Medical Genetics* **3**: 239–57.

Slack J, Nevin N C. (1968) Hyperlipidaemic xanthomatosis: I. Increased risk of death from ischaemic heart disease in first degree relatives of 53 patients with essential hyperlipidaemia and xanthomatosis. *Journal of Medical Genetics* **5**: 4–8.

Slack J, Noble N, Meade T W, North W R. (1977) Lipid and lipoprotein concentrations in 1604 men and women in working populations in north-west London. *British Medical Journal* **ii**: 353–7.

Smith D W. (1966) Dysmorphology (teratology). *Journal of Pediatrics* **69**: 1150–69.

Smith D W. (1977) An approach to clinical dysmorphology. *Journal of Pediatrics* **91**: 690–2.

Smith M. (n.d., *c.* 1999) *Lionel Sharples Penrose: A biography.* Colchester: Michael Smith, privately printed by Lavenham Press, Lavenham, Suffok.

Spencer K, Macri J N, Anderson R W, Aitken D A, Berry E, Crossley J A, Wood P J, Coombes E J, Stroud M, Worthington D J, Doran J, Barbour H, Wilmot R. (1993) Dual analyte immunoassay in neural tube defect and Down's syndrome screening: results of a multicentre clinical trial. *Annals of Clinical Biochemistry* **30**: 394–401.

Stevenson A C. (1953) Muscular dystrophy in Northern Ireland. I. An account of the condition in 51 families. *Annals of Eugenics* **18**: 50–93.

Stevenson A C. (1954) Recording genetic information concerning individuals and families. *Ulster Medical Journal* **23**: 54–60.

Stevenson A C, Davison B C C, Oakes M W. (1970) *Genetic Counselling.* London: William Heinemann Medical Books.

Stevenson A C, Johnston H A, Stewart M I P, Golding D R. (1966) Congenital malformations: a report of a study of series of consecutive births in 24 centres. *Bulletin of the World Health Organization* **34** (Suppl.): 9–127.

Strachan T, Read A P. (1996) *Human Molecular Genetics.* Oxford; New York, NY: BIOS Scientific; Chichester: Wiley-Liss (4th edn, New York, NY: Garland Science, 2010).

Swales J D. (ed.) (1985) *Platt versus Pickering: An episode in recent medical history.* London: Keynes Press for the British Medical Association.

Tansey E M, Christie D A. (eds) (2000) *Looking at the Unborn: Historical aspects of obstetric ultrasound.* Wellcome Witnesses to Twentieth Century Medicine, vol. 5. London: The Wellcome Trust. Freely available online at www.ucl. ac.uk/histmed/publications/wellcome_witnesses_c20th_med

Tassabehji M, Read A P, Newton VE, Harris R, Balling R, Gruss P, Strachan T. (1992) Waardenburg's syndrome patients have mutations in the human homologue of the *Pax-3* paired box gene. *Nature* **355**: 635–6.

Taylor A J, Lloyd J. (1995) The role of the Gene Therapy Advisory Committee in the oversight of gene therapy research in the UK. *Biologicals* **23**: 37–8.

Thein S L, Weatherall D J. (1987) Approach to the diagnosis of beta-thalassaemia by DNA analysis. *Acta Haematologica* **78**: 159–67.

Thomson A L. (1975) *Half a Century of Medical Research. 2: The Programme of the Medical Research Council (UK).* London: MRC.

Thompson A, Taylor N. (eds) (2006) *Hamlet.* London: Arden Shakespeare.

Trotter W. (1916) *Instincts of the Herd in Peace and War 1916–19.* London: Fisher Unwin.

Turner G, Turner B, Collins E. (1970) Letter: Renpenning's syndrome: X-linked mental retardation. *Lancet* **296**: 365–6.

Turpin R, Bernyer G. (1947) De l'influence d l'hérédité sur la formule d'Arneth (cas particulier du mongolisme). *Revue d'Hématologie* **2**: 189–206.

Tyler A, Ball D, Craufurd D. (1992) Presymptomatic testing for Huntington's disease in the United Kingdom. The UK Huntington's Disease Prediction Consortium. *British Medical Journal* **304**: 1593–6.

University College London, Library. (1979) *A list of the Papers and Correspondence of Lionel Sharples Penrose (1898–1972): held in the manuscripts room, UCL Library,* compiled by M Merrington. London: Publications Office, UCL.

Wald N J, Watt H C, Haddow J E, Knight G J. (1998) Screening for Down syndrome at 14 weeks of pregnancy. *Prenatal Diagnosis* **18**: 291–3.

Wald N J, Kennard A, Densem J W, Cuckle H S, Chard T, Butler L. (1992) Antenatal maternal serum screening for Down's syndrome: results of a demonstration project. *British Medical Journal* **305**: 391–4.

Wald N J, Rodeck C, Hackshaw A K, Walters J, Chitty L, Mackinson A M. (2003) First and second trimester antenatal screening for Down's syndrome: the results of the Serum, Urine and Ultrasound Screening Study (SURUSS). *Health Technology Assessment* **7**: 1–77.

Watt D C. (1998) Lionel Penrose FRS (1898–1972) and eugenics: Parts one and two. *Notes and Records of the Royal Society of London* **52**: 137–51; 339–54.

Watt D C. (2000) C P Blacker, R A Fisher and L Penrose on eugenic fundamentals, Part I. *Galton Institute Newsletter* No. 37 (June) freely available at www.galtoninstitute.org.uk/Newsletters/GINL0006/eugenic_fundamentals.htm (visited 19 January 2010).

Weatherall D J. (1985) *The New Genetics and Clinical Practice.* Oxford: Oxford University Press.

Weatherall D J. (2001) Historical review: towards molecular medicine; reminiscences of the haemoglobin field, 1960–2000. *British Journal of Haematology* **115**: 729–38.

Weatherall D J. (2002) Sir Cyril Astley Clarke. *Biographical Memoirs of Fellows of the Royal Society* **48**: 69–85.

Weatherall D J, Old J M, Thein S L, Wainscoat J S, Clegg J B. (1985) Prenatal diagnosis of the common haemoglobin disorders. *Journal of Medical Genetics* **22**: 422–30.

Weber W W, Mittwoch U, Delhanty J D. (1965) Leucocyte alkaline phosphatase in Klinefelter's syndrome. *Journal of Medical Genetics* **39**: 112–15.

Wells R S, Kerr C B. (1965) Genetic classification of ichthyosis. *Archives of Dermatology* **92**: 1–6.

Wilkinson L. (1997) Sir Austin Bradford Hill: medical statistics and the quantitative approach to prevention of disease. *Addiction* **92**: 657–66.

Winter R M, Baraitser M, Douglas J M. (1984) A computerised data base for the diagnosis of rare dysmorphic syndromes. *Journal of Medical Genetics* **21**: 121–3.

Wolstenholme G. (ed.) (1984) Julia Bell. *Munk's Roll* 7: 31–2.

Wolstenholme G. (ed.) (1989) Cedric Oswald Carter. *Munk's Roll* **8**: 78–80.

Woodrow J C. (1975) HL-A and its association with clinical disease. HL-A associations in clinical research. *Proceedings of the Royal Society of Medicine* **68**: 802–4.

Woodrow J C. (1985) Genetic aspects of the spondyloarthropathies. *Clinics in Rheumatic Diseases* **11**: 1–24.

Woodrow J C. (1988) Genetics of the spondyloarthropathies. *Baillière's Clinical Rheumatology* **2**: 603–22.

World Federation of Neurology, Research Committee, Research Group on Huntington's chorea. (1989) Ethical issues policy statement on Huntington's disease molecular genetics predictive test. *Journal of the Neurological Sciences* **94**: 327–32.

World Federation of Neurology, Research Group on Huntington's Disease. (1993) Presymptomatic testing for Huntington's disease: a world-wide survey. *Journal of Medical Genetics* **30**: 1020–2.

Yates F, Mather K. (1963) Ronald Aylmer Fisher. *Biographical Memoirs of Fellows of the Royal Society* **9**: 91–130.

Yates J R, Malcolm S, Read A P. (1989) Guidelines for DNA banking. Report of the Clinical Genetics Society working party on DNA banking. *Journal of Medical Genetics* **26**: 245–50.

Zallen D T, Christie D A, Tansey E M. (eds) (2004) *The Rhesus Factor and Disease Prevention.* Wellcome Witnesses to Twentieth Century Medicine, vol. 22. London: The Wellcome Trust Centre for the History of Medicine at UCL. Freely available online at www.ucl.ac.uk/histmed/publications/wellcome_witnesses_c20th_med

Biographical notes[*]

Professor Sir Donald Acheson
KBE FRCP FFPH FFOM FRCS FRCOG HonFRSM (1926–2010) trained first as an internist at the University of Oxford and the Middlesex Hospital, and subsequently specialized in public health. He was an acting squadron leader at the RAF medical board (1953–55), a travelling fellow of University College, Oxford (1957–59), then medical tutor at the Nuffield department of medicine at the Radcliffe Infirmary and in 1968 he began a long association with Southampton University, where he was professor of clinical epidemiology until 1983 and foundation dean of the faculty of medicine. He was Chief Medical Officer of England (1983–91), dealing with the HIV/AIDS crisis. His principal research contributions have been to discover the relationship between lack of exposure to UV light as a cause of multiple sclerosis and the association of cancer of the ethmoid sinus to inhalation of wood and leather dust in industrial workers.

Ms Chris Barnes
MSc RMN RGN (b. 1955) had a background in mental health nursing and nursing research before becoming a genetic counsellor at the genetics centre at Guy's Hospital in 1988. She has been actively involved in the UK Association of Genetic Nurses and Counsellors (AGNC) since 1990, and is a member of both the genetic counsellor training panel and the genetic counsellor statutory regulation steering group.

Professor Sir John Bell
Kt FRCP HonFREng FRS PMedSci (b. 1952) educated at the University of Alberta, he qualified in medicine at Oxford (Rhodes and Commonwealth scholar), with house jobs at John Radcliffe Hospital, Oxford, the department of clinical cardiology, Hammersmith Hospital, London; the renal unit, Guy's Hospital, London and in neurology at Queen Square, London, becoming a research fellow at the Nuffield department of clinical medicine, Oxford. He was a clinical fellow, department of medicine and postdoctoral fellow, department

---

* Contributors are asked to supply details; other entries are compiled from conventional biographical sources.

of medical microbiology at Stanford University (1982–87) and a Wellcome senior clinical fellow and honorary consultant physician, Nuffield department of clinical medicine and surgery, John Radcliffe Hospital (1987–89). He was appointed university lecturer, Oxford (1989–92), Nuffield professor of clinical medicine (1992–2001) and Regius professor of medicine, University of Oxford (2001–). He has been chairman of the Office for Strategic Co-ordination of Health Research (OSCHR) and the Oxford Health Alliance; founder of the Wellcome Trust centre for human genetics and a founder fellow of the Academy of Medical Sciences and has been president since 2006. He is also a member of the board of the UK Biobank Ltd and a trustee of the Rhodes Trust, Nuffield Medical Trust and the Ewelme Almshouse charity. His research has contributed to a clearer understanding of the genetic determinants of susceptibility in type 1 diabetes and rheumatoid arthritis and he has helped pioneer a large number of high-throughput genomic methodologies applied to biomedical science.

Dr Julia Bell

FRCP (1879–1979) studied mathematics at Cambridge University before becoming statistical assistant to Karl Pearson at the Galton Laboratory for National Eugenics. She assumed principal responsibility for the *Treasury of Human Eugenics*, published between 1909 and 1958. She trained in medicine during the First World War (1914–20) and was a Medical Research Council research assistant (1920–65). In 1944 she was made an honorary research associate at UCL. See Wolstenholme (ed.) (1984); Bundey (1996); Harper (2005).

Dr Caroline Berry

PhD FRCP (b. 1937) qualified in 1961 and trained at Middlesex Hospital, London; was medical officer to the paediatric research unit at Guy's Hospital from 1974, becoming a senior registrar in the South-east Thames Regional Genetics Centre in 1976. She was appointed consultant clinical geneticist in 1979 and clinical director in 1993. She served on the council of the Clinical Genetics Society and chaired the medical study group of the Christian Medical Fellowship.

Professor Martin Bobrow

CBE FRCP FRCPath FRS FMedSci (b. 1938) came to Britain after graduating in South Africa. He worked in Edinburgh and Oxford and held chairs of medical genetics in Amsterdam and

Guy's Hospital before becoming professor of medical genetics in Cambridge in 1995. He has been on the council of the MRC, a governor of the Wellcome Trust, national chairman of the Muscular Dystrophy Campaign, and chairman of the Clinical Genetics Society. He has also held posts as chairman of COMARE (Committee on Radiation in the Environment), ULTRA (Unrelated Living Transplant Regulating Authority), deputy chairman of the Nuffield Council on Bioethics and a member of the Human Genetics Advisory Commission.

Professor Sir John Burn
Kt MD FRCP FRCPCH FRCOG FMedSci (b. 1952) completed an intercalated genetics degree in 1973 and qualified in Newcastle in 1976. After medical and paediatric training he became clinical scientific officer in the MRC clinical genetics unit, Great Ormond Street Hospital, London, and returned to Newcastle as their first consultant clinical geneticist in 1984. He became first clinical director of the Northern Genetics Service (1989–2004) and professor of clinical genetics, Newcastle University (1991–) and head of the Institute of Human Genetics (2004–10). He was president of the European Society of Human Genetics (2007), chair elect (2011) of the British Society for Human Genetics, chair of the National Institute for Health Research genetics specialty group, Leeds (2008–) and director of the NIHR collaborative group on genetics in healthcare, Leeds (2009–). He has been lead clinician, NHS North East, since 2009.

Professor Stuart Campbell
DSc FRCPEd FRCOG FACOG (b. 1936) graduated from the faculty of medicine, University of Glasgow, in 1961 and in 1965 worked as a research registrar under Professor Ian Donald at the Queen Mother's Hospital, Glasgow. In 1968 he took up a lectureship in obstetrics and gynaecology at the Queen Charlotte's Maternity Hospital in London, working under Professor Sir John Dewhurst, and became senior lecturer in 1973 and professor of clinical obstetrics and gynaecology in 1976. He was professor of obstetrics and gynaecology at King's College Hospital, London (1976–96) and has been professor and chair at the department of obstetrics and gynaecology and the fetal medicine unit at St George's, University of London, since 1996. He was president of the International Society of Ultrasound in Obstetrics and Gynaecology (1990–98) and is an honorary fellow of the American Institute of Ultrasound in Medicine

and honorary life member of the British Medical Ultrasound Society.

Professor Cedric Carter

FRCP (1917–84) was director of the MRC genetics unit, Institute of Child Health (1964–82). He founded the UK Clinical Genetics Society in 1972 and, with John Fraser Roberts, conducted the first genetic counselling clinic in Britain at the Hospital for Sick Children, Great Ormond Street, London. See Wolstenholme (ed.) (1989); Figure 7.

Dr Ian Lister Cheese

PhD FRCP FRCPH (b. 1936) read natural sciences followed by training in research, then trained in medicine, qualifying in 1966. After posts as medical registrar at the Radcliffe Infirmary, he entered general practice in Wantage, Oxfordshire, where he became a tutor in general practice, a trainer in the vocational training scheme, also serving in NHS management in Oxfordshire. In 1984 he entered the senior civil service and held appointments in the Department of Health and the Department of Education. His posts included responsibilities for the fitness of teachers, hospital services for children and for genetics services. He was secretary to the Standing Medical Advisory Committee and to the Gene Therapy Advisory Committee. He also served on the RCP Clinical Genetics Committee. Following notional retirement in 1996 he became an adviser to the Department of Health on matters relating to clinical governance and the working of the Abortion Act, has undertaken policy work for the RCP and the Academy of Medical Royal Colleges, was a member of the editorial board that prepared the new formulary, *Medicines for Children*, and served as trustee to voluntary bodies concerned with the support of disabled children and their families.

Professor Angus Clarke

DM FRCP FRCPCH (b. 1954) studied medical and natural sciences at Cambridge, taking his Part II in genetics, and then qualified in medicine from Oxford University. After registration, he worked in general medicine and then paediatrics. He studied the clinical and molecular genetic aspects of ectodermal dysplasia in Cardiff and then worked in clinical genetics and paediatric neurology in Newcastle upon Tyne, developing an interest in Rett syndrome and neuromuscular disorders. He returned to Cardiff in 1989 as senior lecturer in clinical genetics and has been professor in clinical genetics since 2000. He has maintained his interests in Rett syndrome and ectodermal dysplasia and has developed further interests in genetic screening, the genetic counselling process and the social and ethical issues around human genetics. He represents the CMO for Wales on the UK Human Genetics Commission. He has (co)authored and edited six books, including *Genetics, Society and Clinical Practice* (1997) and *Living with the Genome* (2006). He established and directs the Cardiff MSc course in genetic counselling.

Professor Sir Cyril Clarke

KBE FRCP FRCOG FRS (1907–2000) became consultant physician to the Liverpool teaching hospitals, was appointed professor of medicine at the Liverpool Medical School in 1963 and established the Nuffield Institute of Medical Genetics, University of Liverpool, which he directed (1963–72). He was professor and honorary Nuffield research fellow in the department of genetics at the University of Liverpool (1966–76), later emeritus. He served as president of the Royal College of Physicians (1972–77) and was awarded the Lasker clinical research award for work on the genetics of Rh factor and haemolytic disease of the newborn, held jointly with Dr Ronald Finn, Dr John Gorman, Dr Vincent Freda and Dr William Pollack in 1980. See Weatherall (2002).

Professor John Clegg

HonFRCP FRS (b. 1936) was on the Medical Research Council senior scientific staff (1979–2001) and has been professor of molecular medicine at the University of Oxford since 1996.

Dr Gerald Corney

MD FRCOG (1928–2001) qualified at the University of Liverpool and trained in obstetrics and child heath. Following a short-term commission in the Royal Navy as surgeon lieutenant, some of which was spent in the Far East on a survey ship, he was a member of scientific staff at the MRC human biochemical genetics unit at the Galton Laboratory at UCL, where he remained until his retirement in 1989 when he was made an honorary lecturer in the department of genetics and biometry, University of London. His interest in twinning began with his work with Percy Nylander on the Aberdeen registry of twin births and his MD thesis was on twin–twin transfusion in monozygotic twins. See MacGillivray (2002).

Dr Clare Davison

MD D(Obst) RCOG DPH (b. 1934) graduated at Queen's University, Belfast. Following hospital appointments, general practice and obtaining DPH in 1962 she went to the MRC population genetics research unit in Oxford and stayed until 1970 when she married and went to Cambridge. In 1974 she was appointed consultant clinical geneticist at Addenbrooke's Hospital, Cambridge, and was responsible for setting up a regional genetics service in East Anglia. She retired in 1994 and in 1996 spent three months as locum clinical geneticist in Dublin, Eire.

Professor Joy Delhanty

PhD FRCPath FRCOG (b. 1937) graduated in zoology (with special emphasis on genetics) at University College London, University of London, gaining a doctorate in human genetics at UCL. She worked at the Galton Laboratory, UCL, under Lionel Penrose in the pioneering days of human cytogenetics. She has held academic posts at UCL from 1961, becoming professor of human genetics in 1998, until her retirement in 2003, later emeritus. She directed the clinical cytogenetics unit of University College Hospital (1994–99) and has been director of the UCL centre for preimplantation genetic diagnosis since 1997. See Figure 4.

Dr Nick Dennis

FRCP (b. 1944) trained in medicine at Cambridge and St Thomas' Hospital, London. He

was a member of scientific staff, MRC clinical genetics unit, Institute of Child Health, London, under Professor Cedric Carter (1972–76) and senior lecturer and honorary consultant in clinical genetics at Southampton (1978–2007).

Professor Dian Donnai

CBE FRCP FRCOG(ad eundem) FMedSci (b. 1945) qualified at St Mary's Hospital Medical School, University of London and trained in paediatrics at St Mary's Hospital, Northwick Park Hospital and Sheffield. In 1978 she took up one of the first senior registrar training posts in medical genetics at St Mary's Hospital, Manchester, and became a consultant there in 1980. She was appointed honorary professor of medical genetics in the University of Manchester in 1994 and then to a substantive chair in 2001. She was president of the Clinical Genetics Society (1997–99), consultant advisor to the CMO (1998–2004) and president of the European Society of Human Genetics (2009–10). In 2005 she was appointed CBE for services to medicine.

Professor Alan Emery

MD PhD DSc FRCP FRCPE FACMG FLS FRSA FRS(E) (b. 1928) after military service, he graduated BSc in biological sciences and subsequently took a medical degree at Manchester University in 1960. After junior hospital appointments, he gained a PhD at Johns Hopkins University in genetics. Returning to the UK in 1964, he set up a small medical genetics unit at Manchester and was appointed foundation professor of human genetics at Edinburgh University in 1968. He has made a life-long study of muscular dystrophy: the disease Emery–Dreifuss muscular dystrophy and its defective protein product, emerin, are both named after him. Subsequently he established the European Neuromuscular Centre (ENMC) based first in Paris then in the Netherlands (1989–99) and was its chief scientific advisor from 1999. He established the Royal Society of Medicine's Section of Medical Genetics and was its first president (2001–04). He has been a research fellow and later an honorary fellow of Green Templeton College, University of Oxford since 1985.

Sir Ronald Aylmer Fisher

Kt FRS (1890–1962) was Galton professor of eugenics at UCL in 1933, and Arthur Balfour professor of genetics at the University of Cambridge (1943–57). See Yates and Mather (1963).

Dr Charles Ford

FRS (1912–99) was head of the cytogenetics section at the Medical Research Council radiobiology unit, Harwell (1949–71) and a member of the Medical Research Council's external staff at the Sir William Dunn School of Pathology, University of Oxford (1971–78). In 1959 he was responsible for some of the first studies of human chromosome abnormalities. See Lyon (2001).

Professor George Fraser

MD PhD DSc FRCP FRCPC (b. 1932) qualified in medicine at Cambridge, followed by a PhD from the University of London and fellowships at the Canadian College of Medical Genetics and the American College of Medical Genetics. He was appointed scientific officer, MRC population genetics research unit, Oxford (1959–61); research fellow, division of medical genetics, University of Washington, Seattle (1961–63); lecturer, department of research in ophthalmology, Royal College of Surgeons, London (1963–66); reader in genetics, University of Adelaide, Australia (1966–68); associate professor, division of medical genetics, University of Washington (1968–71); professor of human genetics, University of Leiden, Netherlands (1971–73); professor of medical genetics, consultant in clinical genetics, Memorial University, St John's, Newfoundland, Canada (1973–76); chief of department of congenital anomalies and inherited diseases, department of national health and welfare, Federal Government of Canada, Ottawa, Canada (1976–79); associate professor, centre for human genetics, McGill University, Montreal, Quebec, Canada (1979–80); special expert in human genetics, National Library of Medicine, National Institutes of Health, Bethesda, Maryland; attached to Moore clinic for medical genetics, Johns Hopkins University, Baltimore, Maryland (1980–84); senior clinical research fellow, Imperial Cancer Research Fund, honorary consultant in clinical genetics, cancer genetic clinic, Churchill Hospital, Oxford (1984–97) until his retirement in 1997. He has been honorary professor of human genetics, Institute of Child Health, UCL, since 2007.

Dr John Fraser Roberts

CBE DSc FRCP FRCPsych FRS (1899–1987) began research in human genetics in Edinburgh during the 1930s. After the Second World War, he was honorary consultant in medical genetics at the Royal Eastern Counties Hospital, Colchester (1946–57) and started a genetics clinic in 1946 at the Hospital for Sick Children, Great Ormond Street, London, the first in Europe (1946–64) and later at the Children's Hospital, Bristol. The MRC created a clinical genetics research unit at the Institute of Child Health, Great Ormond Street, in 1957 with Fraser Roberts as director, where he remained until his retirement in 1964. See Pembrey (1987); Hill (1988); Polani (1987). See Figure 6.

Mrs Margaret Fraser Roberts

(b. 1927) joined the junior staff of the department of epidemiology and medical statistics at the London School of Hygiene and Tropical Medicine in 1947 to be trained as a computer in medical statistics. She was appointed personal assistant/computer to Dr John Fraser Roberts in March 1948 and worked with him for the next 33 years. In 1957 she moved with him to the clinical genetics unit at Great Ormond Street Hospital, London, until Dr Fraser Roberts' retirement in 1964 and then accompanied him to the paediatric research unit at Guy's Hospital, London (1964–81). See Hill (1988).

Professor John Burdon Sanderson Haldane

FRS (1892–1964) was professor of genetics at UCL (1933–37), Weldon professor of biometry there (1937–57) and head of the biometry, genetics and eugenics department of which the Galton Laboratory was part. In addition to numerous fundamental contributions to basic genetics, he was also much involved in early human genetics discoveries, including the first discovery of genetic linkage in man and the first estimate of mutation rate for a human gene. See Clark (1968); Pirie (1966); Kevles (2007).

Professor David Harnden
FRCPath FRSE (b. 1932) was a scientific member of the Medical Research Council radiobiology unit, Harwell (1957–59) and a scientific member of the Medical Research Council clinical and population cytogenetics unit, Edinburgh (1959–69). He was professor of cancer studies at the University of Birmingham (1969–83) and honorary professor of experimental oncology at the University of Manchester (1983–97) and director of the Paterson Institute for Cancer Research there. He was chairman of the South Manchester University Hospitals NHS Trust (1997–2002).

Professor Peter Harper
Kt FRCP (b. 1939) graduated from Oxford University in 1961, qualifying in medicine in 1964. After a series of clinical posts, he trained in medical genetics at the Liverpool Institute for Medical Genetics under Cyril Clarke and at Johns Hopkins University, Baltimore, under Victor McKusick. He was professor of medical genetics at the University of Wales College of Medicine, Cardiff, until his retirement in 2004, later university research professor in human genetics (emeritus), Cardiff University. He has been closely involved with the identification of the genes underlying Huntington's disease and muscular dystrophies, and with their application to predictive genetic testing. He has also been responsible for the development of a general medical genetics service for Wales and has a particular interest in the historical aspects of human and medical genetics. See Harper (1995).

Professor Harry Harris
FRCP FRS (1919–94), biochemist and geneticist, qualified in medicine at Trinity College, Cambridge, served in the forces and joined the Galton Laboratory at UCL in 1947 where he pioneered the field of human biochemical genetics, became a lecturer in the department of biochemistry (1950–53), senior lecturer (1953–58), reader in biochemical genetics (1958–60) and professor of biochemistry, University of London

(1960–65). He was honorary director of the MRC human biochemical genetics research unit (1961–76), professor of human genetics, University of London (1965–76); and Harnwell professor of human genetics, University of Pennsylvania (1976–90), later emeritus. See Hopkinson (1994); Harris (1959).

Dr Hilary Harris

FRCGP (b. 1943) qualified in medicine at the University of Liverpool, and was a principal in general practice, firstly in Liverpool, then for 27 years as senior partner and GP trainer in south Manchester. She ran a Wolfson Foundation trial on cystic fibrosis screening in early pregnancy in primary care. Following this she was appointed to the Advisory Committee on Genetic Testing and then as a member of the UK Government. Human Genetic Commission (1999–2004). See Harris *et al.* (1993). See Figure 14.

Professor Rodney Harris

BSc MD FRCP DTMH FRCPath (b. 1932) graduated in anatomy with physical anthropology at the University of Liverpool in 1956. He trained as a doctor in Liverpool in 1959 before working as the Darwin research fellow of the Eugenics Society in Nigeria in 1960 and in South West Africa (now Namibia) in 1961. He subsequently held clinical and academic posts at Liverpool Royal Infirmary and the University of Liverpool before becoming professor of medical genetics and honorary consultant physician at Manchester Royal Infirmary and the University of Manchester. He was consultant advisor in medical genetics to several Chief Medical Officers at the DHSS, chairman of the Royal College of Physicians committee on clinical genetics, president of the Clinical Genetics Society and chairman of the European Concerted Action on Genetic Services. He was appointed CBE in 1996 and has been emeritus professor of medical genetics in Manchester since 1997. See Figure 14.

Sir Austin Bradford Hill

FRS (1897–1991), medical statistician, was professor of medical statistics at the London School of Hygiene and Tropical Medicine (1945–61). A series of 17 articles published by him in the *Lancet* in 1937 introduced the medical researcher to the use of statistics, later reprinted as Hill (1937). His first attempts to introduce the concept of randomization in controlled trials were for the Medical Research Council. See Wilkinson (1997).

Professor Shirley Hodgson
DM D(Obst)RCOG DCH
FRCP (b. 1945), daughter of
Lionel Penrose: avoided working
in genetics for many years, but
after training in medicine and
working in general practice while
her children were young, she did a
locum in clinical genetics at Guy's
Hospital, and found it irresistible.
She went on to work in the field
of clinical genetics for many years,
promoting the development of
cancer genetics clinics at Guys, St
Mark's and St George's hospitals in
London in the 1990s and ran the
regional cancer genetics service at
Guy's. She has published widely on
the subject, and co-authored several
books. She has an active research
programme investigating inherited
aspects of cancer predisposition,
particular in breast and colorectal
cancers. She is particularly
interested in international
collaborative research. She took up
a new post in cancer genetics at St
George's, University of London, in
2003. Her current research looks
at molecular changes in colorectal
polyps in relation to inherited
colorectal cancer susceptibility, and
she was principal investigator of a
five-year randomized study funded
by Cancer Research-UK to evaluate
whether the Mirena intrauterine
progestagen-releasing system
reduces the risk of endometrial
cancer in women at increased risk
of this condition (with Lynch
syndrome).

Professor Patricia Jacobs
OBE FRSE FRCPath FRCPE
FRCOG FRS (b. 1934) was
a Medical Research Council
scientist (1957–72), professor
in the department of anatomy
and reproductive biology at the
University of Hawaii School of
Medicine (1988–2001), professor
and chief of the division of human
genetics in the department of
paediatrics at Cornell University
Medical College (1985–87) and
director of the Wessex Regional
Genetics Laboratory (1988–2001).

Dr Alan William Johnston
MD FRCP FRCP(Ed) FRCPG
(b. 1928) qualified in medicine
at University College Hospital
Medical School, trained with Victor
McKusick as a fellow in medicine at
Johns Hopkins Hospital (1959/60),
returned to complete an MD
at Cambridge and was resident
assistant physician at University
College Hospital Medical School
(1962–66). He was appointed
consultant physician, Aberdeen
teaching hospitals (1966–92) and
clinical senior lecturer in medicine
and genetics until his retirement
in 1992. He was president,
Clinical Genetics Society (1987/8);
president, Scottish Society of
Physicians (1988). He has been a
member of the Christian Medical

Fellowship since 1950 and an elder in the Church of Scotland since 1977.

Mrs Ann Kershaw

MSt BA RGN RHV (b. 1949) trained as a nurse at Guy's Hospital, London, qualifying in 1972. She worked in the community in Sheffield, and then trained and subsequently worked as a health visitor and clinical teacher in Cambridge (1985–89). She joined the East Anglian medical genetics department at Addenbrooke's Hospital, Cambridge, in 1989, was appointed senior genetic nurse counsellor in 1995 and a consultant genetic counsellor in 2008. She was a member of the genetic counsellor registration working party and is a former section editor for the *BSHG News* and has recently joined their editorial board. She has a long interest and commitment in working with families with Huntington disease and sits on two working groups for the European Huntington Disease Network (EURO HD). She is a member of the United Kingdom Genetic Testing Network (UKGTN) working party on systems and communications.

Mrs Lauren Kerzin-Storrar

MS (b. 1955) studied biology and genetics at University of California, San Diego before qualifying as a genetic counsellor at the University of California, Berkeley and San Francisco in 1979. She was the first Master's trained genetic counsellor to be appointed in the UK in 1981 at St Mary's Hospital, Manchester, where she has continued to work (as consultant and head genetic counsellor since 1993). She has been director of the MSc genetic counselling programme at University of Manchester since it was established in 1992 as the first training programme for non-medical genetic counsellors in Europe. Through clinical practice, teaching, research and professional body contributions she has promoted the integration of psychosocial considerations alongside genetic science in the training and practice of both doctors and genetic counsellors.

Professor Michael Laurence

DSc FRCPE FRCPath (b. 1924) qualified at Liverpool and trained in Wales and was career registrar in pathology at Portsmouth (1953–55), research fellow in hydrocephalus and spina bifida at the Hospital for Sick Children, Great Ormond Street, London (1955–58). He was appointed senior lecturer in paediatric pathology at the Welsh National School of Medicine (1959–69), reader in applied genetics (1969–76) and professor of paediatric research (1976–89), later emeritus. He was project leader, Euroregister of Congenital Malformations, for Wales (1985) and published over 300 scientific papers on congenital malformations, clinical genetics, paediatric pathology, hydrocephalus and spina bifida, prenatal diagnosis and prevention of malformations. He was a member of the Clinical Genetics Society, later secretary and president; of the British Paediatric Association, the Society for Research into Hydrocephalus and Spina Bifida (past secretary and president). See Rocker and Laurence (eds) (1981).

Professor Victor McKusick

(1921–2008) qualified in medicine at Johns Hopkins University and completed his internship and residency in internal medicine there. He was executive chief of the cardiovascular unit at Baltimore Marine Hospital (1948–50) while progressing through the ranks in the Johns Hopkins department of medicine. He also held joint professorships in epidemiology in the Johns Hopkins University school of public health and in biology at Johns Hopkins University. He founded the division of medical genetics in 1957, which he headed until 1973, when he became the William Osler professor and chairman of the department of medicine and physician in chief of

Johns Hopkins Hospital. He held these posts until 1985, when he was named university professor of medical genetics.

Professor Ursula Mittwoch

DSc (b. 1924) entered the Galton Laboratory as a student for a doctorate in 1947 and later postdoctoral appointments to work on the biochemistry of cystinuria (in collaboration with Harry Harris and Bette Robson) and on cytogenetics. During the 1960s she carried out genetic testing by demonstrating muco-polysaccharide inclusions in the lymphocytes of patients with Hurler and Hunter syndromes. Since then she has specialized in the genetics of sex determination with special reference to cell proliferation and energy metabolism. She retired as professor of genetics in 1989, later emeritus. See Mittwoch (1995).

Professor Bernadette Modell

FRCP FRCOG (b. 1935) started her academic life in 1952 as a biologist with a particular interest in genetics, embryology and anthropology. She studied for her PhD in Cambridge when the science of molecular biology was in its earliest exciting stages. She was professor of community genetics in the department of obstetrics and gynaecology and then the department of primary care and population sciences at the Royal Free and University College Medical School, London (1993–2000), later emeritus in the department of health informatics and multiprofessional education (CHIME). She began her work on thalassaemia at UCL and University College Hospital in 1964, pioneering the treatment, and prenatal diagnosis of the disease, and then with WHO and others to develop the global concepts of community genetics. She was a Wellcome principal research fellow (1992–2000).

Professor Michael Modell
FRCP FRCGP DCH (b. 1937) qualified at University College Hospital and University College Medical School and worked as a partner in a large multi-ethnic general practice (James Wigg Practice; Kentish Town Health Centre, London, 1963–93). The main focus of his academic work has been on undergraduate and postgraduate medical education and primary care genetic research. He was the acting head and professor of the UCL department of primary care and population sciences (2000–02) until his retirement, later emeritus. In recent years he has represented the Royal College of General Practitioners on the fetal anomaly screening

programme steering group and the Genetics Commissioning Advisory Group (GenCAG).

Sir Alan Moncrieff
KBE OBE MD FRCP HonFRCOG (1901–71) qualified at the Middlesex Hospital and joined the staff of the Hospital for Sick Children as house physician in 1925. He studied in Paris and then in Hamburg and returned to become the medical registrar and pathologist to Great Ormond Street Hospital. In 1945 he was appointed the first Nuffield professor of child health at the University of London until his retirement in 1964, later emeritus. He was medical correspondent to *The Times* and author and editor of many books. In 1961 he was the first to receive the James Spence Medal of the British Paediatric Association, and was knighted in 1964. See Norman (1983).

Professor Marcus Pembrey
FRCP FRCPCH FRCOG FMedSci (b. 1943) trained in medical genetics in Liverpool (1969–71) and Guy's Hospital, London (1973–78). In 1979 he moved to the Institute of Child Health, London, as head of the mothercare unit of paediatric genetics where he led a team that helped to introduce DNA testing into clinical genetics in the 1980s. He was also

consultant clinical geneticist at the Hospital for Sick Children, Great Ormond Street, London (1979–98) and consultant adviser in genetics to the Chief Medical Officer (CMO), Department of Health (1989–98). After early retirement in 1998, he continued as visiting professor and director of genetics within the Avon Longitudinal Study of Parents and Children, University of Bristol, until 2006. See Pembrey (2004).

Professor Lionel Penrose
FRCP FRS (1898–1972) was resident director of the Royal Eastern Counties Institution, Colchester (1930–39). He was Galton professor of eugenics (1945–62) and of human genetics (1962–65) at University College London. See Harris (1973); Figure 1.

Sir Robert Platt (Lord Platt of Grindleford from 1967)
Bt FRCP (1900–78) was professor of medicine at the University of Manchester (1945–65) and a member of the Royal Commission on medical education (1965–68). He was president of the Royal College of Physicians (1957–62). See Platt (1972).

Professor Paul Polani
FRCP HonFRCP(Ire) FRCPCH FRCOG HonFRCPath DCH FRS (1914–2006), geneticist born in Trieste, who showed that Down syndrome and Klinefelter syndrome were both caused by the presence of an extra chromosome, and Turner syndrome by a missing chromosome. He qualified at Siena and Pisa, became a ship's surgeon lieutenant on a British merchant vessel and was interned as an enemy alien on the Isle of Man when Italy joined the war in 1940. He was released in 1941 and appointed medical and surgical officer at the Evelina Children's Hospital, Southwark, London. In 1948 he was appointed assistant to the director, department of child health, Guy's Hospital Medical School, London (1950–55), director, paediatric research unit (renamed the division of genetics and development, King's College, London; 1960–83); director, medical research unit, National Spastic Society (1955–60); Prince Philip professor of paediatric research, University of London (1960–80), later emeritus, and honorary paediatrician, Guy's Hospital and Medical School (1960–85). He was also director, South East Thames Regional genetics centre (1976–82), a fellow of King's College London, and geneticist, division of genetics and development at Guy's Hospital. See Richmond (2006); Adinolfi and Alberman (2006); Harper (2007). See Figure 8.

Professor Sue Povey

MD FMedSci (b. 1942) graduated in natural sciences (genetics) at Cambridge in 1964 and qualified in medicine in 1967. After brief clinical experience at University College Hospital, London, Huddersfield and working for Save the Children Fund in Algeria, she returned to UCL to join the MRC human biochemical genetics unit under Harry Harris in 1970. Having obtained an MD in 1977, she was deputy director of the unit (1989–2000) and was appointed Haldane professor of human genetics at UCL and editor of the *Annals of Human Genetics* until her retirement in 2007. Her interest in tuberous sclerosis continues in the curation of the *TSC1* and *TSC2* locus-specific mutation databases and she has chaired a working group drafting ethical guidelines for such databases.

Dr Robert Race

CBE FRS (1907–84) was director of the Medical Research Council blood group unit (1946–73). See Clarke (1985).

Professor Andrew Read

MA PhD FRCPath FMedSci (b. 1939) trained in organic chemistry in Cambridge and did his PhD on RNA chemistry in Sir Alexander Todd's department there. After postdoctoral work in Heidelberg and Warwick, he joined the department of medical genetics at the University of Manchester, in 1977 as lecturer, subsequently senior lecturer, reader and, in 1995, professor of human genetics. His early work was on the diagnosis of neural tube defects and their prevention by periconceptional vitamin supplementation. In 1982 he established one of the first DNA diagnostic laboratories in the UK, now one of the two national genetics reference laboratories. He was a founder member, and subsequently chairman, of the Clinical Molecular Genetics Society and later founder chairman of the British Society for Human Genetics. His research focused on mapping and identifying the genes responsible for Mendelian conditions, including *PAX3*, the first homeodomain gene found to cause a human genetic condition, type 1 Waardenburg syndrome. He is co-author of two of the leading textbooks in the area (Strachan and Read (1996); Read and Donnai (2007)).

Professor James Renwick

(1926–94), worked initially at the Galton Laboratory, UCL, London, and subsequently at University of Glasgow. He was responsible for some of the earliest human genetic linkage studies, as well as for developing computerized approaches to genetic linkage analyses. His later work, again at UCL, was on the teratological basis of congenital malformations.

Professor Derek Roberts

FRSE (b. 1925) was professor of human genetics at the University of Newcastle upon Tyne until his retirement in 1990, later emeritus; honorary consultant to the Royal Victoria Infirmary, Newcastle upon Tyne; and honorary director, Genetic Advisory Service NHS Northern Region (1965–90).

Professor Elizabeth (Bette) Robson

PhD (b. 1928) obtained her PhD with Penrose and Haldane in 1954. After a Rockefeller fellowship at Columbia University, New York, in 1954/5 she worked with Harry Harris at the London Hospital Medical College and then King's College London (1955–62) where, together with Oliver Smithies and David Hopkinson, they developed pioneering methods for the detection of enzymes after starch gel electrophoresis. In 1962, as a founder member of the MRC human biochemical genetics unit, she moved back to UCL, succeeding Harry Harris as Galton professor of human genetics (1978–93). In 1969 she mapped the Haptoglobin locus to chromosome 16 by linkage analysis, one of the earliest autosomal assignments. With a younger colleague, the late Dr Peter Cook, she continued to play a major role in the application of protein polymorphisms and blood groups in linkage analysis,

being a founder member of the Human Genome Organization and a great supporter of the human gene mapping meetings. She was joint editor of the *Annals of Human Genetics* (1978–93). See Figure 4.

Professor Heather Skirton

PhD MSc RGN (b. 1953), a qualified midwife and registered genetic counsellor, trained as a nurse and midwife in Melbourne, Australia, emigrated to the UK in 1988 and worked as a genetic counsellor (1989–2000) and nurse consultant in genetics (2000–04), during which time she completed a PhD on the outcomes of genetic counselling (1996–99). Her academic posts include appointments as co-director of the MSc in genetic counselling at Cardiff University (2000–03) and reader in health genetics at the University of Plymouth (2004–08). She was promoted to professor of applied health genetics at the University of Plymouth in 2008. She chaired the Joint Committee on Medical Genetics (2003–05), established in 1999 and representing the Royal College of Pathologists (RCPath), the Royal College of Physicians (RCP) and the British Society for Human Genetics (BSHG) to provide a unified forum. She is past president of the International Society of Nurses in Genetics (2006).

Dr Alan Stevenson

CBE FRCP (1909–95) was the director of the Medical Research Council population genetics unit, Oxford, and lecturer in human genetics at the University of Oxford (1958–74). See Gillam and Macdonald (eds) (2001).

Professor E M (Tilli) Tansey

PhD PhD HonFRCP FMedSci (b. 1953) is convenor of the History of Twentieth Century Medicine Group and professor of the history of modern medical sciences at the Wellcome Trust Centre for the History of Medicine at UCL.

Lord Walton of Detchant

TD MD DSc FRCP HonFRCPE FRCPath FRCPsych FRCPCH FMedSci (b. 1922) qualified at King's College Medical School, trained at the Universities of Durham and Newcastle upon Tyne; served in the Royal Army Medical Corps, commanded the Territorial Army's I(N) General Hospital (1963–66), becoming an honorary Colonel (1968–73). He worked at the Massachusetts General Hospital and Harvard University; was appointed as a consultant neurologist, University of Newcastle Hospitals (1958–83) where he studied muscular dystrophy, building up an extensive research team at the Newcastle General Hospital in the muscular dystrophy group research laboratories, discovering a new classification based on genetic information. He was professor of neurology, University of Newcastle upon Tyne (1968–83), dean of medicine there (1971–81) and warden of Green College, Oxford (1983–89). He was a member of the MRC (1974–78) and was a member of the House of Lords' Select Committee on science and technology (1991–97). He was chairman of the Hamlyn national commission on education (1991–93); a member of the General Medical Council (1971–90), chairman of the education committee (1975–82), and president (1982–89); president of the British Medical Association (1980–82), Association for the Study of Medical Education (1982–94), Royal Society of Medicine (1984–86), Association of British Neurologists (1987/8); first vice-president of the World Federation of Neurology (1981–89), president (1989–97), and has been life president of the Muscular Dystrophy Group GB since 2000 and chairman (1970–95). See Rose (1992).

Professor Sir David Weatherall

Kt DL FRCP FRCPE FRS (b. 1933) qualified at Liverpool University in 1956 and was professor of haematology at the University of Liverpool (1971–74), Nuffield professor of clinical medicine at the University of Oxford (1974–92), Regius professor of medicine at the University of Oxford (1992–2000); honorary director of the MRC molecular haematology unit (1980–2000) and the institute for molecular medicine, Oxford (1988–2000, renamed the Weatherall Institute of Molecular Medicine from 2000), later emeritus. His major research contributions have been in the elucidation of the clinical and molecular basis for the thalassaemias and the application of this information for the control and prevention of these diseases in the developing countries. He has been Chancellor of Keele University since 2002.

Professor Robin Winter

(1950–2004) geneticist and dysmorphologist, qualified at University College Hospital Medical School and took an intercalated degree in genetics at the Galton Laboratory. After various house jobs, he was a visiting fellow at the department of human genetics, Richmond, Virginia, where he studied under Walter Nance. On his return he took up a senior registrar post, and was appointed clinical geneticist at the Kennedy-Galton Centre at Northwick Park in 1981, where he and Michael Baraitser began their work on the London dysmorphology and neurogenetics databases (Winter *et al.* (1984)). In 1992 he moved to the Institute of Child Health and Great Ormond Street Hospital for Children NHS Trust, and was appointed to a personal chair in clinical genetics and dysmorphology there from 1994 until his death in 2004, aged 53, after a short illness with cancer of the oesophagus. He was an active member of the council of the Clinical Genetics Society and served as president (2001–03). See Baraitser (2004); Nance (2004); Rosser and Wilson (2004); Figure 12.

Professor John Woodrow

FRCP (b. 1924) was on the staff of the department of medicine, University of Liverpool (1961–91), and was consultant physician in general medicine and rheumatology to the Liverpool United Hospitals. He formed part of the research team at the Liverpool Institute for Medical Genetics.

Glossary*

alphafetoprotein
One of the principal plasma proteins present in the fetus, whose altered level in maternal blood and amniotic fluid may be useful in the prenatal diagnosis of **neural tube defects** and **Down syndrome**.

amniocentesis
The technique of removal of fluid from the pregnant uterus for genetic and other analyses.

Barr body
See **sex chromatin**

Bayesian risks
An approach to risk estimation based on the combination of prior and modified risks, first proposed by the eighteenth-century philosopher and cleric, Thomas Bayes.

β-thalassaemia
A severe form of anaemia resulting from failure of red blood cells to produce adequate amounts of the haemoglobin β chain.

coarctation of the aorta
A congenital heart malformation, seen more often in males and also in the chromosome disorder **Turner syndrome**.

creatine kinase
An enzyme present in muscle. An elevated level is found in the blood of those affected by many types of muscle damage, including those affected by or carrying several types of muscular dystrophy. This can be useful in their diagnosis.

cytogenetics
The study of chromosome structure and function.

DNA probes
Lengths of DNA sequence labelled so as to allow their corresponding sequence to be identified by hybridization for diagnosis or research.

Down syndrome
One of the most common human chromosome abnormalities, caused by presence of an extra copy of chromosome 21. Named after Langdon Down, the original describer.

drumstick
A darkly staining nuclear appendage, representing a second X chromosome in **polymorphonuclear leucocytes**, first described by Davidson and Smith (1954).

*Terms in bold appear in the Glossary as separate entries. Compiled by Professors Angus Clarke, Peter Harper, Bernadette Modell, Heather Skirton, Sir David Weatherall and Dr Caroline Berry.

Duchenne muscular dystrophy
The most common form of childhood **muscular dystrophy**, determined by a gene on the X chromosome and occurring mainly in young boys.

dysmorphology
The study of human malformation syndromes.

epiloia
A previous name for **tuberous sclerosis**.

eugenics
The study of ways in which the overall genetic constitution of human populations can be 'improved'.

genetic linkage
The occurrence of genes close together on the same chromosome, resulting in co-inheritance of characteristics more often than expected by chance.

haemoglobinopathies
A group of inherited anaemias, including sickle cell disease and **thalassaemias,** resulting from abnormality in the structure or production of haemoglobin.

Huntington's disease
A progressive brain disease, following autosomal dominant inheritance, resulting in abnormal movements, with motor and mental deterioration.

ichthyosis
A scaly appearance of the skin usually caused by a mutation in one of the many structural proteins expressed in the skin (especially the keratin proteins and filaggrin). The associated conditions are known collectively as the ichthyoses.

karyotype
The overall chromosome complement of a cell or individual.

Klinefelter syndrome
A condition of infertility and hypogonadism in males, resulting from an XXY chromosome constitution.

Leopard syndrome
An inherited condition characterized by multiple pigmented lesions (lentigines), deafness and cardiac defects.

molecular genetic tests
Tests based on analysis of DNA.

mongolism
A previous name for **Down syndrome**.

monozygotic
Twins derived from the same fertilized egg.

muscular dystrophy
A group of inherited and progressive neuromuscular disorders of childhood and later life. *See* **myotonic dystrophy, Duchenne muscular dystrophy.**

myotonic dystrophy
The most common form of adult **muscular dystrophy**, characterized also by numerous non-muscular features.

neural tube defects
Developmental disorders of the nervous system, including spina bifida and anencephaly, caused by incomplete closure of its primitive tube structure.

pharmacogenetics
The scientific field of genetic variation in the response to or metabolism of drugs.

phenylketonuria
An inherited metabolic disorder resulting in accumulation of the amino acid phenylalanine, and causing mental handicap if not detected and treated early.

polymorphonuclear leucocyte
One form of white blood cell, particularly responsible for combating bacterial infection.

preimplantation genetic diagnosis (PGD)
The diagnosis of genetic conditions by analysis of cells taken from the embryo prior to its implantation in the uterus.

Renpenning syndrome
A form of X-linked mental handicap.

Rhesus haemolytic disease
Anaemia of a Rhesus positive fetus in a Rhesus negative mother, caused by destruction of fetal red blood cells by maternal antibodies to the Rh blood group substance on the red blood cells of the fetus while in the uterus.

Robertsonian translocation
A form of chromosome abnormality resulting from fusion of the short arms of particular chromosome pairs, such as of chromosome 21 and chromosome 14 in rare cases of **Down syndrome.** See Figure 3.

Rubinstein Taybi syndrome
A rare developmental syndrome with characteristic facial and other features.

sex chromatin (Barr) body
A body seen microscopically under the nuclear membrane in most female mammalian cells, first described by Murray Barr and Ewart Bertram. It represents the inactivated member of the pair of X chromosomes in a female mammalian cell.

triploidy
Presence of a complete additional set of chromosomes and therefore giving a chromosome count of 69.

trisomy
Presence of an additional copy of a specific chromosome.

tuberous sclerosis
An inherited neurodevelopmental disorder causing mental handicap, epilepsy, tumours and skin pigmentary changes, among other features.

Turner syndrome
A condition of infertility and gonadal maldevelopment in females, commonly with heart defects, and due to an XO chromosome constitution and the presence of a single sex chromosome (an X chromosome) instead of the usual two sex chromosomes (XX or XY).

tylosis
Severe thickening (hyperkeratosis) of the palms of the hands and the soles of the feet, which is associated with cancer of the oesophagus and is inherited in a dominant fashion.

Werdnig Hoffman disease
A severe form of spinal muscular atrophy, caused by inherited degeneration of the anterior horn nerve cells of the spinal cord.

X-linked
Determined by a gene located on the X chromosome.

Index: Subject

Index: Names

Biographical notes appear in bold

Lightning Source UK Ltd.
Milton Keynes UK
17 June 2010

155655UK00001B/15/P